Polynomial and Spline Approximation

NATO ADVANCED STUDY INSTITUTES SERIES

Proceedings of the Advanced Study Institute Programme, which aims at the dissemination of advanced knowledge and the formation of contacts among scientists from different countries

The series is published by an international board of publishers in conjunction with NATO Scientific Affairs Division

A Life Sciences **B Physics**	**Plenum Publishing Corporation** **London and New York**
C Mathematical and Physical Sciences	**D. Reidel Publishing Company** Dordrecht, Boston and London
D Behavioral and Social Sciences	**Sijthoff International Publishing Company** **Leiden**
E Applied Sciences	**Noordhoff International Publishing** **Leiden**

Series C – Mathematical and Physical Sciences

Volume 49 – Polynomial and Spline Approximation

Polynomial and Spline Approximation

Theory and Applications

Proceedings of the NATO Advanced Study Institute held at Calgary, Canada, August 26–September 2, 1978

edited by

BADRI N. SAHNEY

University of Calgary, Canada

D. Reidel Publishing Company

Dordrecht : Holland / Boston : U.S.A. / London : England

Published in cooperation with NATO Scientific Affairs Division

Library of Congress Cataloging in Publication Data

CIP

Nato Advanced Study Institute on Polynomial and Spline Approximation,
University of Calgary, 1978.
Polynomial and spline approximation.

(NATO advanced study institutes series : Series C, Mathematical and physical sciences ; v. 49)
Includes index.
1. Approximation theory–Congresses. 2. Spline theory–Congresses.
3. Polynomials–Congresses.
I. Sahney, Badri N. II. Title. III. Series.
QA221.N37 1978 511'.4 79-12247
ISBN 90-277-0984-X

Published by D. Reidel Publishing Company
P.O. Box 17, Dordrecht, Holland

Sold and distributed in the U.S.A., Canada, and Mexico
by D. Reidel Publishing Company, Inc.
Lincoln Building, 160 Old Derby Street, Hingham, Mass. 02043, U.S.A.

Printed in The Netherlands

TABLE OF CONTENTS

PREFACE

The NATO Advanced Study Institute on *Polynomial and Spline Approximation* was held at the University of Calgary between August 26 and September 2, 1978. These Proceedings contain the invited lectures presented to this Institute, the aim of which was to bring together Pure and Applied Mathematicians and possibly Engineers, working in Approximation Theory and Applications. The papers covered a wide area including polynomial approximation, spline approximation, finite differences, partial differential equations and finite element methods.

I am thankful to the International Advisory Committee, whose members were: Professor E.W. Cheney, Professor J. Meinguet, Dr. G.M. Phillips and Professor H. Werner.

I also wish to extend my thanks to Dr. C. Nasim and Dr. K. Salkauskas, the members of the local Organizing Committee. It was through their hard work in the last stages of planning and arrangements that it was possible to organize such a smooth-running ASI. My special gratitude to Dr. Salkauskas who read several of the papers in final manuscript form, and for his helpful suggestions to minimize the errors. To the many secretaries and others who, directly or indirectly worked for the ASI, my heartfelt thanks, and particularly to Mrs Jenny Watkins who worked on this project for almost three years.

The publication of these Proceedings could only be possible with the cooperation of the speakers. The Advanced Study Institute was financed by the NATO Scientific Affairs Division. Modest financial support also came from the National Research Council of Canada and the Dean of the Faculty of Science at the University of Calgary. To all of them I am most thankful.

There were forty-three participants whose names and addresses are included at the end of the book. Last but not least, I wish to express my sincere gratitude to all of them for their contributions in helping to make this ASI a successful and pleasant event.

Calgary, Canada. Badri N. Sahney
November 1978.

SOME APPLICATIONS OF POLYNOMIAL AND SPLINE APPROXIMATION

Lothar Collatz
Hamburg, Germany

Summary
The distinction is described between qualitative error estimations (order of magnitude of the error) and quantitative error bounds (numerically computable strong mathematical pointwise error bounds). Progress by using approximation methods is made in the last years in singular nonlinear boundary value problems, in the method of finite elements, in free boundary value problems (exact inclusion for the free boundary in simple cases) and other areas.

1. Introduction

Approximation methods are today very important for many applications in science, chemistry, biology, economics, etc. Originally the calculation of functions on a computer was a field of application of approximation, but now approximation methods are very useful for linear and nonlinear ordinary and partial differential equations, integral equations and more general functional equations, as they occur frequently in applications. But usually the approximation problems which occur in applications are much more complicated than the problems considered in the classical theory; one has difficulties coming from the fact, that multivariate approximation, singularities, combi-approximations, unusual restrictions, free boundaries etc. occur.

The problems in the applications are often so difficult, that numerical and theoretical analysts have to work together to try

Badri N. Sahney (ed.), Polynomial and Spline Approximation, 1-15.

to solve these problems.

Often the problems are so complex that the only way to obtain acceptable results is to use numerical methods.

If one does not know enough about the type of the wanted function, then it is the natural way to approximate the function by polynomials (in several variables), and if one expects that the value of the function vary strongly, one can divide the considered domain B in smaller parts and one has spline-approximation. Therefore polynomial and spline-approximations are very important for the applications, and furthermore one knows more about these two types than about approximation with other classes of functions. In many cases approximation methods are the only ones which give exact inclusions for the wanted solutions. I intend to give examples for cases in which one can give such exact inclusions and in which this was not yet possible some years ago, for instantce for free boundaries. On the other side these methods are of course not yet powerful enough for the great complexity of many technical problems. Much research is necessary in this area, because only very simple problems are today accessible for a strong mathematical treatment.

2. Extimations and bounds for the error

We consider equations of the form

(2.1) $$Tu = \theta,$$

where u is a wanted element, f.i. in a Banach space (a function or a vector of functions a.o.), T is a given linear or nonlinear operator and θ the zero element in the considered Banach space. Let w be an approximate solution for the solution u and ε the error:

(2.2) $$\varepsilon = w - u .$$

There are two kinds of error considerations:

I) A Qualitative error estimation has the form

(2.3) $$|\varepsilon| = Q(h^p) ;$$

h may be for instance the mesh size in case of finite differences and p the order of accuracy. (2.3) means, that there exists a constant K with

(2.4) $$|\varepsilon| \leq K h^p .$$

K is usually not known.

II) A Quantitative error bound is

(2.5) $$|\varepsilon| \le \gamma \; ;$$

γ is computed, for instance $|\varepsilon| \le \frac{1}{2}\,10^{-5}$. That means: 5 decimals the computer has printed can be guaranteed.

3. Error bounds with aid of the maximum principle

Let us consider a very simple example of a Dirichlet-Problem with discontinuous boundary values. For the temperature $u(x,y)$ in a plate with the doamin

$$B = \{(x,y),\ |x| < 2,\ |y| < 1\} ,$$

we have the Laplace-equation

(3.1) $$\Delta u \equiv \frac{\partial^2 u}{\partial x^2} + \frac{\partial^2 u}{\partial y^2} = 0 \text{ in } B$$

and the boundary values

(3.2) $$u = \begin{cases} 1 \text{ for } |x| = 2 \\ 0 \text{ for } |y| = 1 \end{cases} \quad \text{on } \partial B.$$

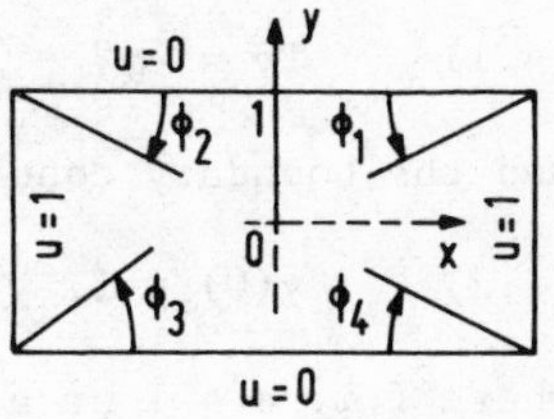

Fig.1

We have singularities at the four corners; we introduce the angles ϕ_j $(j = 1,2,3,4)$ at the four corners, fig. 1, and approximate u by a function w

(3.3) $$u \approx w = \frac{2}{\pi} \sum_{j=1}^{4} \phi_j + \sum_{k=1}^{p} a_k v_k (x,y).$$

Here a_k are real parameters and the v_k are polynomials which satisfy the equation (3.1) and the symmetries of the problem:

(3.4) $$\Delta\phi_j = \Delta v_k = 0 \ (j = 1,2,3,4,\ k = 1,\dots,p).$$

For instance the first of these polynomials are

(3.5) $$v_1 = 1, \quad v_2 = x^2 - y^2, \dots .$$

Now we have to determine the free parameters a_k in such a way, that the boundary conditions (3.2) are satisfied as well as possible. The maximum principle gives error bounds: If $|w-u| \le \delta$ on ∂B, then

(3.6) $$|w-u| \le \delta \text{ holds in } B.$$

We wish to get δ as small as possible, that means we have to carry out Tschebyscheff approximation.

Increasing p and the degree of the polynomials one gets better error bounds:

p	$\|w-u\| \le$
2	0.114
3	0.0020
4	0.00063
5	0.000114

4. Error bounds with aid of monotonicity

For nonlinear problems one has to use monotonicity principles.

Example. A function y(x) may be determined by the Fermi equation

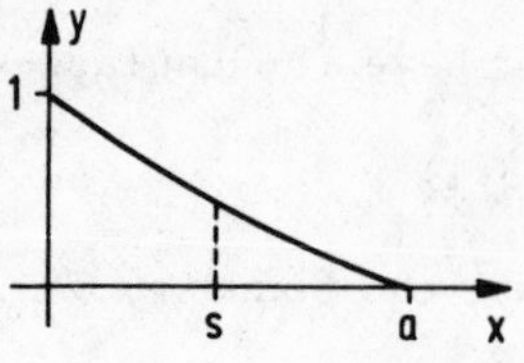

Fig. 2

(4.1) $$Ty = -y'' + \sqrt{\frac{y^3}{x}} = 0$$

and the boundary conditions

(4.2) $$y(0) = 1,\ y(a) = 0,$$

where f.i. $a = 1$ or $a = \infty$. We consider the domain D of function $\phi(x)$ with $\phi(0) = 1$, $\phi(a) = 0$, $\phi > 0$ in $B = \{x \mid 0<x<a\}$. Then the mononicity holds: $[D < C^2(B)]$

(4.3) $$Tv \le Tw \text{ in } B \text{ implies } v \le w \text{ for } v,w \in D.$$

At first we take $a = 1$. We substitute y by the approximate solution w with polynomials in $\sqrt{x}$:

(4.4) $$y \approx w(x) = 1 + a_1x + a_2x^{3/2} + a_3x^2 + \cdots = 1 + \sum_{\nu=1}^{p} a_\nu x^{\frac{\nu+1}{2}}.$$

We have the onesided Tschebyscheff-Approximation: We determine parameters a_ν such that $Tw \le 0$ in $0 < x < a$ and other parameters $\hat{a}_\nu$ with $T\hat{w} \ge 0$ and get upper and lower bounds; f.i. for $p = 3$, we get $|w-y| \le 0.0004$. For $a = \infty$ we have to use spline-approximation

$$y \approx x(x) = \begin{cases} w_1(x) = 1 + ax + bx^{3/2} & \text{for } 0 \le x \le s \\ w_2(x) = \alpha x^{-k} & \text{for } s \le x \le \infty . \end{cases}$$

For instance, for

$$a = -2.27,\ b = 1.336,\ \alpha = 0.06636,\ s = 1,\ k = 4,$$

we get the lower bound:

$$w(x) \leq y(x) \ .$$

Many different types of boundary value problems in differential equations, of integral- and other equations are treated numerically on computers; we can here only select some few simple examples.

5. Integral equations

We mention briefly recent work of Sprekels [78] and Voss [77]. For the equation

$$\text{(5.1)} \qquad Tu(t) = \int_0^1 k(t,s,\lambda)f(\lambda,x)u(s)ds, \text{ with } \lambda = u(0),$$

we ask for a nonnegative nontrivial fixed point; aids are the theorem of Krasnoselskij for expanding operators and the coneiteration, Voss [77]. To make the cone as small as possible is an Approximation problem.

Example. I thank Dr. Voss for the example

$$\text{(5.2)} \qquad \begin{aligned} &-[(1+\cos t)u'] = \lambda^2(1+\lambda\alpha t)u, \\ &u'(0) = (u1) + \lambda\beta u'(1) = 0 \ , \end{aligned}$$

α,β are given constants, λ may be the smallest positive eigenvalue; for some pairs α,β the numerical results are

α	β	lower bound for λ	upper bound for λ
0	0	2.0702923	2.0702989
0.5	0	1.8270384	1.8270457
0.5	0.25	1.3515800	1.3515880

the calculation was done without using Green's function. (computer time for every case circa 15 sec. on TR4 computer).

6. Finite elements

We consider an elliptic differential equation for a function $u(x) = u\ (x_1,\ldots,x_m)$ in an open connected bounded domain B in the m-dimensional point space R^m:

$$\text{(6.1)} \qquad Tu = 0.$$

The boundary B may consist of the parts Γ_j (j=1,2,3) with the j-th boundary condition on Γ_j; we have the first boundary condition with prescribed values of u; analogously on T_2 $\frac{\partial u}{\partial n}$ is prescribed as a second boundary condition, where n is the outer normal. B is divided into subdomains $B_1, B_2, \ldots, B_q$, fig. 3. We have at the interface Γ_{ij} between B_i and B_j with the inner normals v_i and v_j the jumping operator S_{ij}:

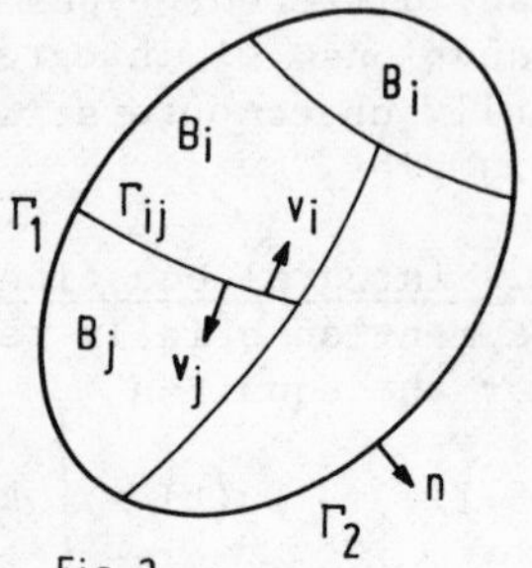

Fig. 3

(6.2) $$S_{ij}u = \frac{\partial u_i}{\partial v_j} - \frac{\partial u_j}{\partial v_j}$$

where u_i is the restriction of the function $u(x)$ on the domain B_i. (geometrical interpretation: Collatz [74])

Then under weak conditions the vector

$$\phi = \{Tu \text{ in } B,\ u \text{ on } \Gamma_1,\ \frac{\partial u}{\partial n} \text{ on } \Gamma_2,\ S_{ij}u \text{ on } \Gamma_{ij}\}$$

is of monotone type, that means (B. Werner [75], Natterer [78], Meyn [78])

$$\phi v < \phi w \text{ implies } v \le w \text{ in } B.$$

I thank Mr. Meyn for a numerical example of reactor physics, fig. 4, for which T-10 cubic finite elements were used, and the error bound was found with the mesh size

$h = \frac{1}{6}$ for the approximate solution w:

$$|u - w| \le 0.00464,$$

f.i. $|0.86306 - u(0,1)| \le 0.00464$.

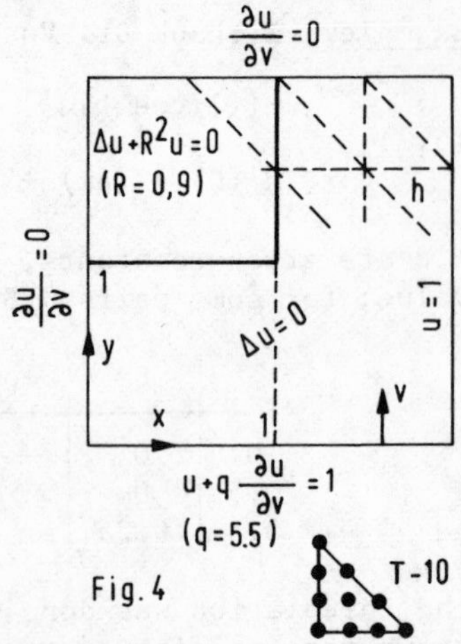

Fig. 4

7. Error bounds for free boundary values problems

There exists a large literature on free boundary value problems, on the Stefan problem a.o. Hoffmann [78] used a nonlinear integral equation for the free boundary in the Stefan problem. Numerical error bounds were given by Collatz [77], [78]. The following methods for getting an exact error bound use only the monotonicity and not the method of integral equations.

The problem of melting ice, f.i. of a lake, may be described by the heat-conduction equation:

(7.1) $$Lu = \frac{\partial u}{\partial t} - \frac{\partial^2 u}{\partial x^2} = 0 \text{ in } B\ (0 < x < s(t),\ t > 0), \text{ fig. 5}$$

and the boundary conditions:

(7.2) $u = h(t)$ for $t > 0$ with $h(0) = 0$, (boundary part Γ)

(7.3) $u = 0,\ \frac{\partial u}{\partial x} = -\frac{ds}{dt}$ for $x = s(t)$.

The "free boundary" $x = s(t)$ is wanted and we will include $s(t)$ in bounds; we consider the case $h(t) = t$ and take $v(x,t)$ as approximation for $u(x,t)$ and the curve $v = 0$ or $x = \hat{s}(t)$ as approximation for $s(t)$:

(7.4) $v = -ax + a^2t$ with $a > 0$

then $v = 0$ is equivalent to $x = \hat{s}(t) = at$.

The monotonicity gives:

(7.5) $\left.\begin{array}{l} Lv \geq 0 \text{ in } B,\ v \geq h(t) \\ \text{implies } \hat{s}(t) \geq s(t) \end{array}\right\}$ on Γ.

We have

$$Lv = a^2 > 0,\ v_\Gamma = a^2t \geq t$$

for $a > 1$, f.i. $a = 1$;

therefore

$$s(t) \leq \hat{s}(t) = t.$$

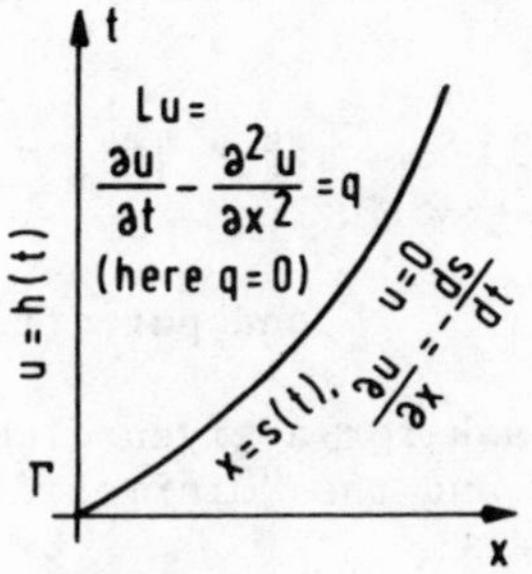

Fig.5

We get a lower bound with

$$v = -ax + a^2t + c(-ax + a^2t)^2 \text{ (constants } a,c).$$

We have

$$v_\Gamma \leq t \quad \text{for} \quad a \leq 1,$$

$$Lv = -a^2[(c-1) + 2cax - 2xa^2t] \geq 0$$

for $a = 0.75$.

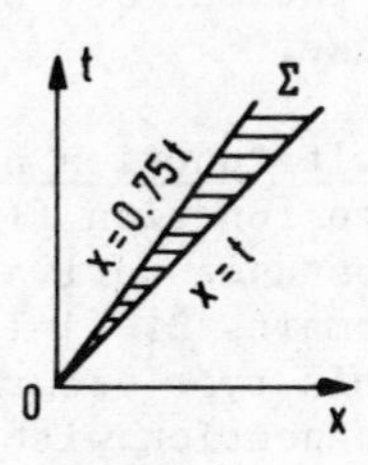

Fig.6

The free boundary is included in the strip Σ, fig. 6.

Of ocurse it is easy to improve these bounds. Many other types of free boundary value problems have been treated numerically, f.i. one can get easily with quadratic polynomials as bounds for $s(t)$, $|0.28875 - s(0.3)| \leq 0.00375$. Here it was only the purpose to illustrate the method and to show how the inverse problem of the free boundary value problem gives inclusions for the free boundary with aid of polynomial approximation.

8. Appendix

The following is only a summary of basic concepts and ideas; the lectures will contain many explanations to more special cases and applications to analysis, differential- and integral equations and other sciences.

I. H-sets in unifunction-approximation Problems.

Definition of unifunction-approximation: Let B be a compact metric space, f.i. a closed bounded connected domain of the n-dimensional point space R^n of points $x = (x_1,\ldots,x_n)$ and C[B] the linear space of continuous functions f(x), defined on B. Let $W = \{W(x,a)\}$ be a subset of C[B], depending on a parameter-vector $a = (a_1,\ldots,a_p)$. Sometimes $p = \infty$ is admitted. All quantities may be real or sometimes complex. We suppose C[B] as metric space with a distance $\rho(f,g)$ for every pair f, $g\varepsilon C[B]$. For many applications we use the classical supremum norm

$$(8.1) \qquad \|f\| = \|f\|_\infty = \sup_{x\varepsilon B} |f(x)|,$$

$$\text{and put } \rho(f,g) = \|f - g\| .$$

We ask for a given element $f(x)\varepsilon C[B]$ for the "nearest" element $\hat{w}\varepsilon W$ and the "minimal distance" ρ_0

$$(8.2) \qquad \rho_0 = \inf_{w\varepsilon W} \rho(f,w), \quad \rho(f,\hat{w}) = \rho_0 .$$

ρ_0 exists in every case, if W is not empty, but a "minimal solution" or "best T-Approximation" w doesnot exist in every case.

Multifunction approximation. If one has to approximate not only the function f(x) but furthermore operators $T_j f (j=1,\ldots,k)$, for instance derivatives, or if one has to approximate in different domains B_j $(j=1,\ldots,k)$, one has multifunction approximation. This type occurs frequently in the applications, also often in connection with inequalities (constraints) (Collatz [69], Bredendiek-Collatz [76]).

Chained approximation. If one or more parameters a_ν occur in the analytic form of W (or in W and in the constraints) at more than one place and if one cannot avoid this phenomenon by a one to one transformation of the parameters, then the approximation is called "chained approximation" (Collatz [75]).

Examples: 1) Approximation of $f(x) = e^x$ in $x \varepsilon [0,1]$ by $w(x) = a_1 + a_2 x + (a_3 + a_3{}^3)x^2$ is not a chained approximation (introduce $a_3{}^* = a_3 + a_3{}^3$),

2) As in 1), but with $w(x) = a_1 + \frac{x}{a_1}$ is a chained approximation.
Let us consider at first the unifunction nonchained approximation.

<u>Fundamental Lemma</u>. D may be a nonempty subset of B with the property: There exist for every $x \varepsilon D$ a (complex) $\varepsilon(x)$ with $|\varepsilon(x)| = 1$ and a real $\alpha(x)$ with: there exists no $w \varepsilon W$ with

(8.3) $\quad \mathrm{Re}[\varepsilon(x)w(x)] < \alpha(x) \qquad$ for all $x \varepsilon D$.

Then we have the inclusion theorem for the minimal distance ρ_0: (compare Meinardus [67])

(8.4) $$\tau = \inf_{x \varepsilon D}\{\alpha(x) - \mathrm{Re}[\varepsilon(x)f(x)]\} \le \rho_0 = \inf_{w \varepsilon W} \rho(f,w).$$

<u>Def.</u>: The domain S is "starshaped" with respect to an element $\hat{s} \varepsilon S$, if for all $s \varepsilon S$ and all $\lambda \varepsilon [0,1]$, fig. 7,

(8.5) $\quad \lambda s + (1-\lambda)\,\hat{s} \;\varepsilon\; S.$

If S is convex, S is starshaped with respect to every $\hat{s} \varepsilon S$.
Then we have the

<u>Theorem</u> (necessary and sufficient condition for a minimal solution):

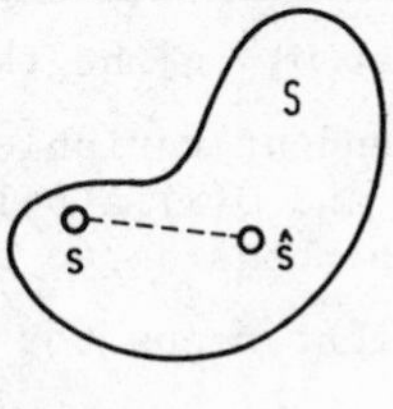

Fig. 7

Given a fixed element $f(x) \;\varepsilon\; C[B]$ and a given element $w(x) \varepsilon W$, where W is starshaped with respect to $\hat{w}$. Let E_w be the set of "extremal points" in B, where the error $\hat{\varepsilon} = \hat{w} - f$ reaches the maximum of the modulus:

(8.6) $\quad E_{\hat{w}} = \{x \varepsilon B,\ |\hat{\varepsilon}(x)| = \|\hat{\varepsilon}\|\}.$

Then $\hat{w}$ is a best T-approximation for f with respect to the class W iff

(8.7) $$\min_{x \varepsilon E_x} \mathrm{Re}[(\hat{W}-f)\overline{(\hat{w}-w)}] \le 0 \quad \text{for all } w \;\varepsilon\; W.$$

Now let us restrict ourselves to the real case.

For the classical cases of linear approximation in one independent variable by a Haar-System and of rational approximation in one independent variable one can use the "alternants" as possible sets D; one uses in other cases, in more general nonlinear and multivariate cases the H-sets instead of alternants.

<u>Definition of H-sets</u>. A set $D = M_1 \cup M_2$ (subset of B) is called an H-set, if there exists no pair w, $\hat{w} \epsilon W$ with

$$(8.8) \qquad w-\tilde{w} \begin{cases} > 0 \text{ in } M_1 \\ < 0 \text{ in } M_2 \end{cases} .$$

For a fixed element $\hat{w} \epsilon W$ let us consider a set $D = M_1 \cup M_2$ (subset of B) with

$$(8.9) \qquad \hat{\varepsilon} = \hat{w} - f \begin{cases} > 0 \text{ in } M_1 \\ < 0 \text{ in } M_2 \end{cases} , \text{ then D}$$

is called an $\hat{H}$-set, if there exists no element w with

$$(8.10) \qquad \hat{w}-w \begin{cases} > 0 \text{ in } M_1 \\ < 0 \text{ in } M_2 \end{cases} .$$

<u>H-sets for rational multivariate Approximation</u>

Let $w[\frac{k}{m},n]$ be the class of rational functions $\frac{P(x)}{Q(x)}$ in n independent variables $x_1,\ldots,x_n$, where P(x) is a polynomial of degree k and Q(x) a polynomial of degree m, and furthermore $Q(x) > 0$ in the considered domain B. Then every H-set for the class $W[\frac{s}{o},n]$ (that means for the polynomials of degree s) is also an H-set for the class $W[\frac{k}{m},n]$, if $s = k + m$. Therefore it is necessary to look only on the H-sets for polynomials.

H-sets for s = 1 (linear polynomials) are given by Taylor [72] for any number n of independent variable, for s = 2 (quadratic polynomials) for n = 2 (Collatz-Krabs [73]) and for n = 3 in special cases.

There exist algorithms in the linear case for testing, whether a given set D of points is an H-set (compare Collatz-Krabs [73]). In the nonlinear case usually only the conditions (8.10) are practicable for this testing and generally not the conditions (8.8).

Examples are given in the lectures.

II. For the applications, important generalisations and multifunction approximation.

Example from economics: $A(t)$ = demand for a certain product at the time t with $0 \leq t \leq t_N$; $P(t)$ = Production of this product at the time t; $P(t)$ is piecewise constant and can be changed at certain time t_j with $0 < t_1 < t_2 < \ldots < t_{N-1}$:

$$(8.11) \qquad p_j = P(t) = \text{const. for } t_{j-1} < t < t_j \ .$$

Every changing causes consts. If at the time t is more produced than sold, it is necessary to put this on a store (stock), the quantity may be $L(t)$; we have

$$(8.12) \qquad L(t) = \int_0^t P(t)dt - \int_0^t A(t)dt.$$

We idealize the phenomenon: We consider

$$(8.13) \qquad \int_0^t A(t)dt = f(t)$$

as given (prescribed, expected), and

$$(8.14) \qquad \int_0^t P(t)dt = w(t)$$

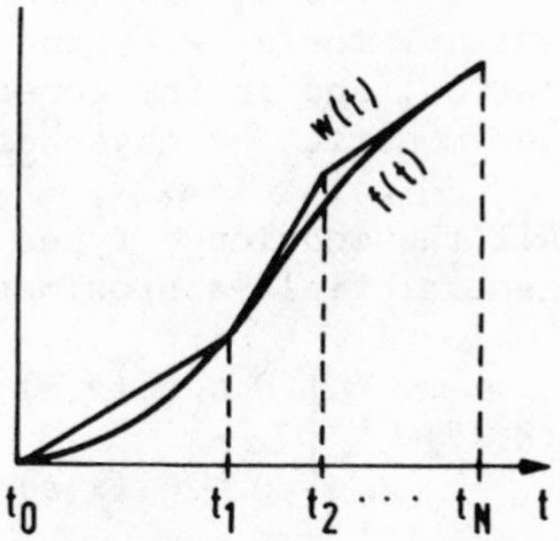

Fig.8

can be influenced by suitable choice of the real numbers t_j, p_j. One wishes to approximate $f(t)$ by $w(t)$ in the following sense, fig. 8:

1) $w(t)$ is piecewise linear; $w(t)$ is a linear spline (Spline-Approximation with free knots; the number of possible knots is limited (given).
2) The store $L(t)$ should never be negative:

$$(8.15) \qquad w(t) \geq f(t) \quad \text{with} \quad L(0) = L(t_N) = 0,$$
(onesided Approximation).

3) One wishes to have the maximum of $L(t)$ as small as possible, the rooms for the store should not be too great (Tschebyscheff Approximation.

Examples from analysis.
Many examples from differential- and integral equations and from the applications are given in the lectures. Frequently we have onesided approximation, which are perhaps more inportant than the

classical-two-sided approximation.
Different types of the multifunction-Approximation (mentioned in I) are:
Syn-Approximation: $T_j f$ $(j=1,\ldots,k)$ are to be approximated in the same domain B.
Simultan Approximation (Bredendiek [69], Bredendiek-Collatz [76]) One has different domains B_j.
Combi-Approximation. One has diferent classes W_j in the different domains.
The classical approximations can be described as optimization problem:

$$(8.16)\qquad \begin{cases} -\delta_1 \le W(x,a) - f(x) \le \delta_2 \text{ for all } x \in B, \\ \delta_1 \ge 0,\ \delta_2 \ge 0 \\ Q = \delta_1 + \delta_2 = \text{infinum}\ . \end{cases}$$

The unknowns are the parameters $a_1,\ldots,a_p$ and δ_1, δ_2.
One has for $\delta_1 = 0$ (or for $\delta_2=0$) onesided Tschebyscheff-Approximation, for $\delta_1 = \delta_2$ the classical twosided Tschebyscheff-Approximation, and in the general case with no restriction on δ_1, δ_2 the unsymmetric Tschebyscheff-Approximation.

All the mentioned types of Approximation can be submitted in the general field-approximation:

$$(8.17)\qquad \begin{cases} 0 \le \phi(x,a) \le \delta, \quad \delta = \text{Min} \\ 0 \le G_j(x,a) \quad (j=1,\ldots,m) \end{cases} \qquad \text{for all } x \in B$$

where ϕ, G_j are given functions of their arguments.

III. Polynomial Approximation in problems with nonlinearities and singularities

a) Nonlinear ordinary differential equation. Bending of a beam with great deflection. Weinitschke [77] considered the boundary value problem for a function y(x) in the interval B=(0,1), fig. 9

$$(8.18)\qquad Ty = -(x^3 y') - \frac{2x^3}{y^2} = 0 \text{ in } B$$

$$(8.19)\qquad Ry = \begin{cases} y'(0) = 0 \\ y(1) - s = 0 \end{cases}$$

with a given constant s, f.i. s=1. We look for an approximate solution w(x) of the form:

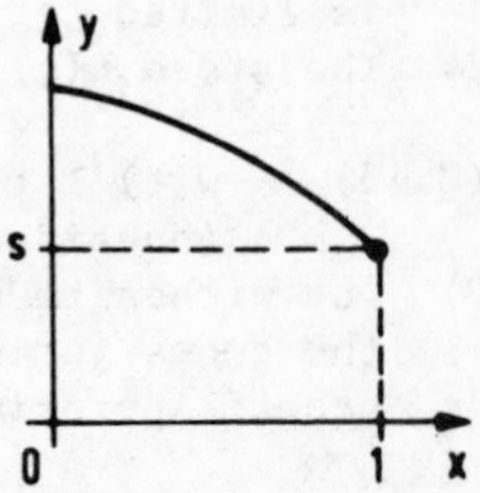

Fig. 9

$$(8.20)\qquad y(x) \approx w(x) = s + \sum_{\nu=2}^{p} a_\nu (1-x^\nu),$$

which satisfies the boundary-conditions (8.19) for every real constant a_ν. Let us take p=3 for simplicity. At first we determine a function $w=w_1$ with parameters a_2, a_3 such that $Tw_1 \leq 0$; then we choose a function $w=w_2$ with other vlaues $\hat{a}_2$, $\hat{a}_3$ of the parameters with $Tw_2 \geq 0$. The relations

$$(8.21) \qquad Tw_1 \leq 0 \leq Tw_2$$

then imply

$$(8.22) \qquad w_1(x) \leq y(x) \leq w_2(x).$$

Here we get

$$|w_1 - w_2| \leq 0.0212$$

and the exact error bound:

$$|y - \frac{w_1+w_2}{2}| \leq 0.0106 .$$

Other examples, for instance with cone-iteration of Voss [77], are given in the lectures.

I thank Mr. Grothkopf for the numerical calculation on a computer.

References

Bredendiek, E. [69] Simultant Approximation, Arch. Rat. Mech. Anal., 33 (1969), 307-330.

Bredendiek, E. and L. Collatz [76] Simultant Approximation bei Randwertaufgaben, to appear in Intern. Ser. Num. Math. 1976.

Cheney E.W. [66] Introduction to approximation theory N.W. McGraw-Hill, 1966.

Collatz, L. [56] Approximation von Funktionen bei einer und bei mehreren unabhängigen Veränderlichen, Z. Angew. Math. Mech. 36 (1956), 198-211.

Collatz, L. [69] Nichtlineare Approximation bei Randwertaufgaben V.JKM. Weimer 1969, 169-182.

Collatz, L. [75a] Methods for Solution of Partial Differential Equations, Symp. Numer. Solution Part. Diff. Equ. Kjeller, Norway 1975, 1-16.

Collatz, L. [74] Monotonicity with discontinuities in partial differential equations, Proc. Conf. Ord. Part. Diff. Equ. Dundee 1974, Lect. Notes Math. Bd. 415 (1974) 85-102.

Collatz, L. [75] Bemerkungen Zur verketteten Approximation, Intern. Ser. Num. Math. 26 (1975), 41-45.

Collatz, L. [77] The numerical treatment of some singular boundary value problems, Conference on Numer. Analysis, Dundee 1977, to appear.

Collatz, L. [78] Application of approximation to some singular boundary value problems, Proc. Conference Num. Anal. Dundee, Springer Lect. Not. Math. Bd. 630 (1978) 41-50.

Collatz, L. - W. Krabs [73] Approximationstheorie, Teubner, Stuttgart, 1973, 208 S.

Hoffmann, K.H. [78] Monotonie bei nichtlinearen Stefan-Problemen, Internat. Ser. Num. Math. 39 (1978), 162-190.

Meinardus, G. [67] Approximation of Functions, Theory and Numerical Methods, Springer Verlag, 1967, 198 S.

Meinardus, G. [76] Periodische Splines Functionen, Lect. Notes Math. Bd. 501 (1976) 177-199.

Meyer, A.G. [60] Schranken für die Lösungen von Randwertaufgaben mit elliptischer Differentialgleichung. Arch. rat. Mech. Anal. 6 (1960), 277-298.

Meyn, K.-H. -B. Werner [78] Randomaximum- und Monotonieprinzipien für elliptische Randwertaufgaben mit Gebietszerlegungen, erscheint demnächst, to appear.

Mitchell, A.R. - R. Wait [77] The finite Element Method in Partial Differential Equations, 1977, 192 p.

Natterer, F. [78] Berechnung oberer Schranken für die Norm der Ritz-Projektion auf finite Elemente Internat. Ser. Num. Math. 39 (1978) 236-245.

Osborne, M.R. - G.A. Watson [78] Nonlinear Approximation Problems in Vektornorms, Lect. Notes Math, 630 Springer 1978, 117-132.

Prenter, P. [71] Polynomial operators and equations, in Proc. Symp. Nonlinear functional analysis and applications, ed. L.B. Rall, Academic Press 1971.

Redheffer, R.M. [63] Die Collatzshce Monotonie bei Anfangswertproblemen, Arch. Rat. Mech. Anal. 14 (1963), 196-212.

Schröder, J. [56] Das Iterationsverfahren bei allgemeinerem Abstandbegriff, Math. Z. 1956, 66, 111-116.

Schröder, J. [59] Vom Defekt ausgehende Fehlerabschätzungen bei Differentialgleichungen. Arch. Rat. Mech. Anal. 3 (1959) 219-228.

Sprekels, J. [78] Iterationsverfahren zur Einschliebung positiver Lösungen superlinearer Integralgleichungen. Intern. Ser. Num. Math. 39, (1978) 261-279.

Taylor, G.D. [72] On minimal H-sets, J. Approximation Theory 5 (1972), 113-117.

Voss, H. [77] Existenz und Einschliebung positiver Losungen superlinearer Integralgleichungen und Randowertaufgaben, Habil. Schrift, Hamburg, 1977.

Walter, W. [70] Differential and Integral Inequalities, Springer 1970, 352 S.

Weinitschke, H. [77] Verzweigungsprobleme bei kreisförmigen elastischen Platten, Intern. Ser. Num. Math. 38 (1977) 195-212.

Werner, B. [75] Monotonie und finite Elemente bei elliptischen Differentialgleichungen, Intern. Ser. Num. Math. 27 (1975) 309-329.

Wetterling, W. [68] Lösungsschranken bei elliptischen Differentialgleichungen, Intern. Ser. Num. Math. 9 (1968) 393-401.

Wetterling, W. [77] Quotienteneinschliebung bei Eignewertaufgaben mit partieller Differentialgleichung, Int. Ser. Num. Math. 38 (1977) 213-218.

SPLINE BLENDED APPROXIMATION OF MULTIVARIATE FUNCTIONS

Charles Hall*
Institute for Computational Mathematics
and Applications
Department of Mathematics and Statistics
University of Pittsburgh

ABSTRACT. Spline and polynomial blended approximants to a multivariate function are synthesized from approximations to univariate samplings of the function. General algebraic and analytic properties of blended and tensor product interpolants are reviewed.

INT. The construction of a polynomial or spline approximant to a function f of more than one variable is typically accomplished by means of <u>tensor product</u> transformations. For example, let

$$a = r_0 < r_1 < \dots < r_N = b, \; c = s_0 < s_1 < \dots < s_M = d$$

be a partition of the rectangular domain $:[a,b]\times[c,d]$, and let $\phi_i(r)$ and $\psi_j(s)$ be functions such that

$$\phi_i(r_k) = \delta_{ik}; \qquad i,k = 0,1,2,\dots,N \tag{1}$$
$$\psi_j(s_k) = \delta_{jk}; \qquad j,k = 0,1,2,\dots,M.$$

Then the <u>bivariate</u> <u>polynomial</u> <u>Lagrange</u> <u>interpolant</u> $L_f(x,y)$ of f is given by

$$L_f(r,s) = \sum_{i=0}^{N} \sum_{j=0}^{M} f(r_i,s_j)\, \phi_i(r)\, \psi_j(s) \tag{2}$$

where $\phi_i(r)$, $i = 0,1,\dots,N$ and $\psi_j(s)$, $j = 0,1,\dots,M$ are chosen as polynomials of degree N and M respectively satisfying (1).

---

*Partially supported by USAF Office of Scientific Research Contract No. F44620-76-0104.

Badri N. Sahney (ed.), Polynomial and Spline Approximation, 17–34.

Another choice for the functions $\phi_i(r)$, $i = 0,1,\ldots, N$ and $\psi_j(s)$, $j = 0,1,\ldots, M$ are the natural cubic splines. There results the natural bicubic spline interpolant $B_f(x,y)$ of f given by

$$B_f(r,s) = \sum_{i=0}^{N} \sum_{j=0}^{M} f(r_i,s_j)\, \phi_i(r)\psi_j(s). \tag{3}$$

Bicubic splines with other endpoint or boundary conditions [4] have similar representations.

We note that the bicubic spline $B_f(r,s)$ (or bipolynomial $L_f(r,s)$) can be generated as the result of applying the transformation

$$\mathcal{P}_r: \; g \to \sum_{i=0}^{N} g(r_i,s)\, \phi_i(r) \tag{4}$$

to the image $\mathcal{P}_s[f]$ under the transformation

$$\mathcal{P}_s: \; d \to \sum_{j=0}^{M} d(r,s_j)\, \psi_j(s) \tag{5}$$

of the function f. That is,

$$B_f(r,s) = \mathcal{P}_r \mathcal{P}_s[f]\,. \tag{6}$$

If we consider the linear spaces

$S_1 \equiv \text{Basis } \{\phi_0(r),\ldots,\phi_N(r)\}$, $S_2 \equiv \text{Basis } \{\psi_0(s),\ldots,\psi_M(s)$

then the tensor product space $S_1 \times S_2$ is defined to be the (N+1)(M+1) dimensional space Basis $\{ \{\phi_i(r)\psi_j(s)\}_{j=0,\ldots,M}^{i=0,\ldots,N}\}$. The function $B_f(r,s)$ (resp. $L_f(r,s)$ belongs to $S_1 \times S_2$ and hence is commonly referred to as the tensor product interpolant of f. Other tensor product approximants (e.g. Galerkin, least squares etc.) can be constructed using the same space $S_1 \times S_2$, or equivalently the same basis.

We next note that the generic transformations $\mathcal{P}_r$ and $\mathcal{P}_s$ given in (4) and (5) are projectors (linear and idempotent). Further, for the choices of ϕ_i and ψ_j

given they are <u>commutative</u> ($\mathcal{P}_r \mathcal{P}_s = \mathcal{P}_s \mathcal{P}_r$). The cardinality conditions (1) imply that

$$\begin{aligned} \mathcal{P}_r[g](r_i,s) &= g(r_i,s),\ 0 \le s \le 1,\ i = 0,1,\ldots,N \\ \mathcal{P}_s[d](r,s_j) &= d(r,s_j),\ 0 \le r \le 1,\ j = 0,1,\ldots,M. \end{aligned} \tag{7}$$

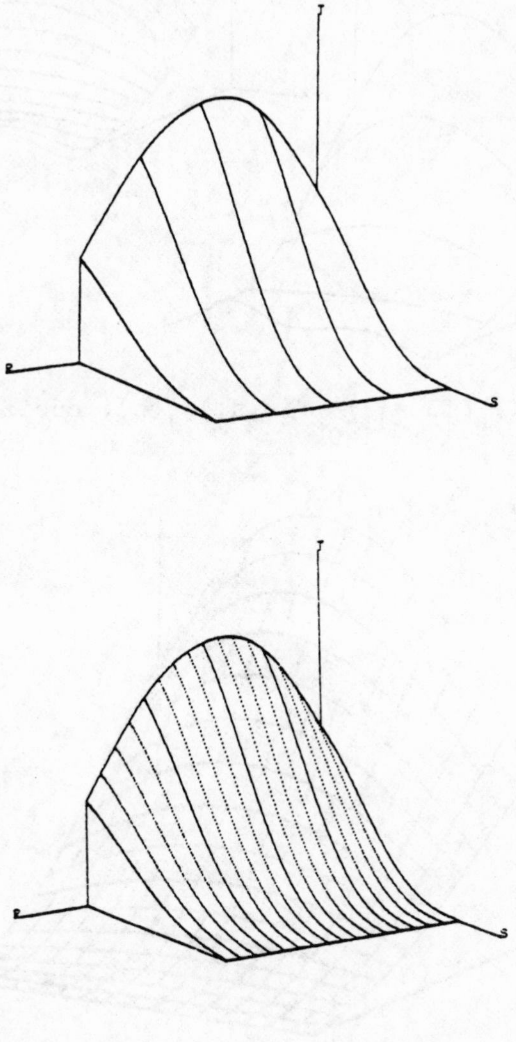

Figure 1. $\mathcal{P}_r[f] = \sum_{i=0}^{4} f(r_i,s)\phi_i(r)$, cubic spline blending.

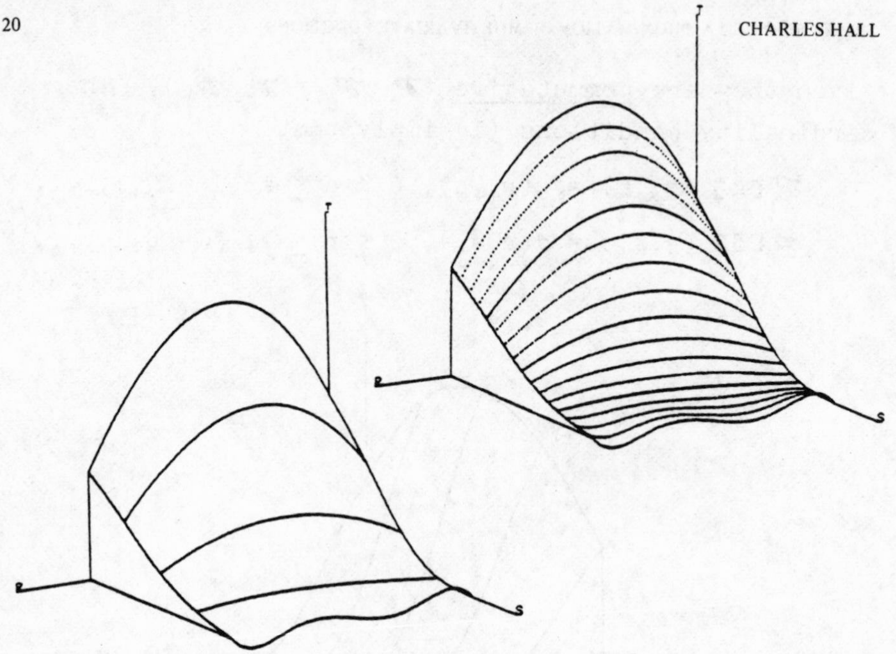

Figure 2. $\mathcal{P}_s[f] = \sum_{j=0}^{4} f(r,s_j)\psi_j(s)$, cubic spline blending.

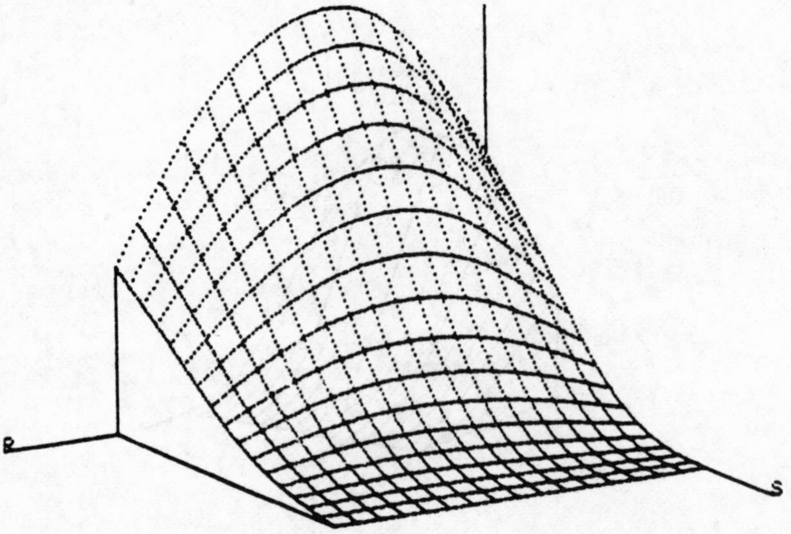

Figure 3. $\mathcal{P}_s\mathcal{P}_r[f] = \mathcal{P}_r\mathcal{P}_s[f]$

Figures 1 and 2 are plots of $\mathcal{P}_r[f]$ and $\mathcal{P}_s[f]$ respectively for a given function f(r,s). $\mathcal{P}_r[f]$ agrees with or interpolates to f along the lines $r = r_i$ $i = 0,1,\ldots, N$ and $\mathcal{P}_s[f]$ agrees with or interpolates to f along the lines $s = s_j$, $j = 0,1,\ldots, M$. Such interpolants which match f on more than a finite point set are termed transfinite [7,8] and the functions $\phi_i(r)$ and $\psi_j(s)$ are called blending functions since they "blend" the given network of curves together to form a surface, [5,6].

A special case of blending involves the choice of $N = 1$, $\phi_0(r) = 1-r$, $\phi_1(r) = r$ in (4). The two curves (0,s,g(0,s) and (1,s,g(1,s)) are then blended together in a process called "railing" (Cf. Figure 4).

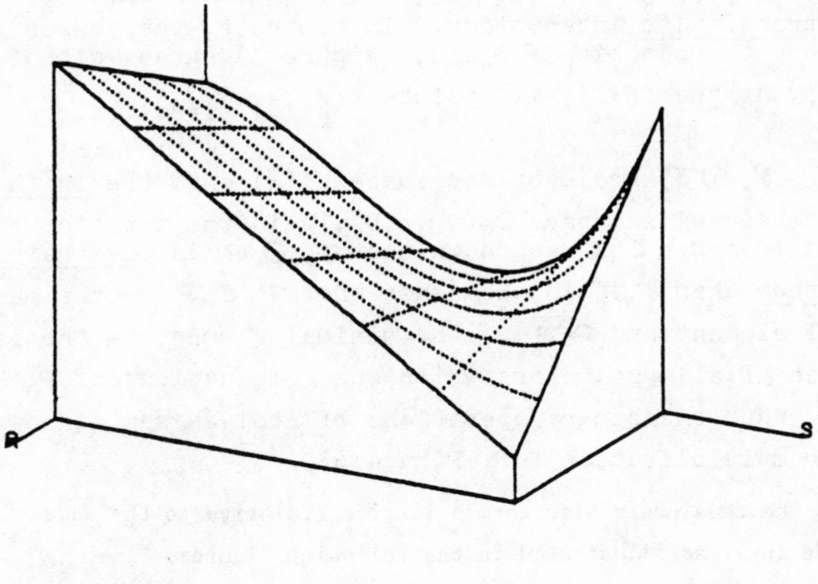

Figure 4. Railing of two curves.

$\mathcal{P}_r \mathcal{P}_s$ projects the space of functions, say, $W_2^1(\mathcal{S})$ onto the finite dimensional subspace $S_1 \times S_2$. Formulae (2) and (3) were "built-up" or synthesized from univariate formulae (4) and (5) as products. Another way to compound the projectors $\mathcal{P}_r$ and $\mathcal{P}_s$, which was studied extensively by Gordon [5,6], makes use of the Boolean sum,

$$\mathcal{P}_r \oplus \mathcal{P}_s : \ f \to \mathcal{P}_r[f] + \mathcal{P}_s[f] - \mathcal{P}_r \mathcal{P}_s[f]. \tag{8}$$

$\mathcal{P}_r \oplus \mathcal{P}_s$ is a projector and the projection $\mathcal{P}_r \oplus \mathcal{P}_s[f]$, (Cf. Figure 5), satisfies

$$\begin{aligned} \mathcal{P}_r \oplus \mathcal{P}_s[f](r_i,s) &= f(r_i,s), \ 0 \le s \le 1, \ i = 0,1,\ldots, N \\ \mathcal{P}_r \oplus \mathcal{P}_s[f](r,s_j) &= f(r,s_j), \ 0 \le r \le 1, \ j = 0,1,\ldots, M. \end{aligned} \tag{9}$$

Hence $\mathcal{P}_r \oplus \mathcal{P}_s[f]$ interpolates to f along lines; it is a transfinite interpolant. In contrast, the tensor product projection $\mathcal{P}_r\mathcal{P}_s[f]$, (Figure 3) agrees with f only at the (N+1)(M+1) points $\{(r_i,s_j)\}_{\substack{i=0,1,\ldots, N.\\ j=0,1,\ldots, M}}$

$\mathcal{P}_r \oplus \mathcal{P}_s$ projects the space $W_2^1(\mathcal{S})$ onto the infinite dimensional subspace $\oplus S(D^2,1,D^2,1)$ in the notation of [3]. Gordon [6] has shown that $\mathcal{P}_r \oplus \mathcal{P}_s$ is algebraically better than $\mathcal{P}_r\mathcal{P}_s$ in the sense that $\mathcal{P}_r \oplus \mathcal{P}_s$ is the maximal element and $\mathcal{P}_r\mathcal{P}_s$ is the minimal element in the lattice of all projectors which are combinations of $\mathcal{P}_r$ and $\mathcal{P}_s$ under the binary operations of Boolean sum and operator multiplication (Cf. Figure 6).

The remainders also form a lattice (relative to the same ordering), as illustrated in the following figures.

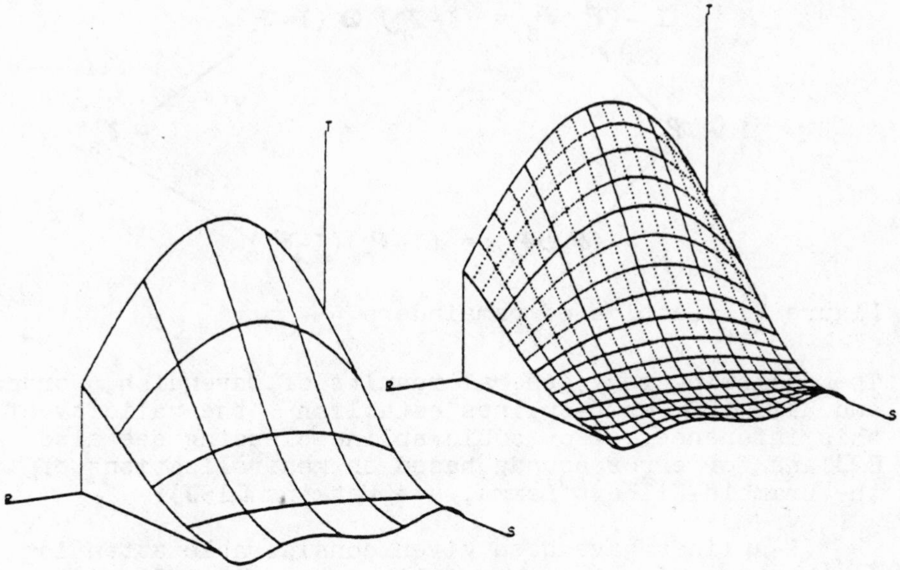

Figure 5. Boolean sum $\mathcal{P}_r \oplus \mathcal{P}_s[f]$.

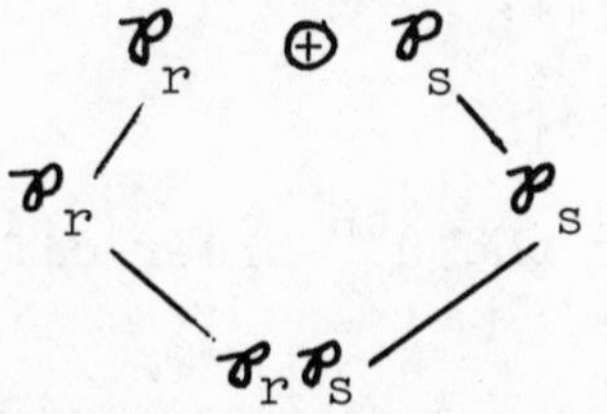

Figure 6. Distributive lattice with partial ordering $A \leq B$ if and only if $A B = A$.

Figure 7 which suggests that the error in the Boolean sum scheme is "smaller" in some sense than for the tensor product scheme. That is, it suggests that <u>analytically</u> $\mathcal{P}_r \oplus \mathcal{P}_s$ is better than $\mathcal{P}_r\mathcal{P}_s$.

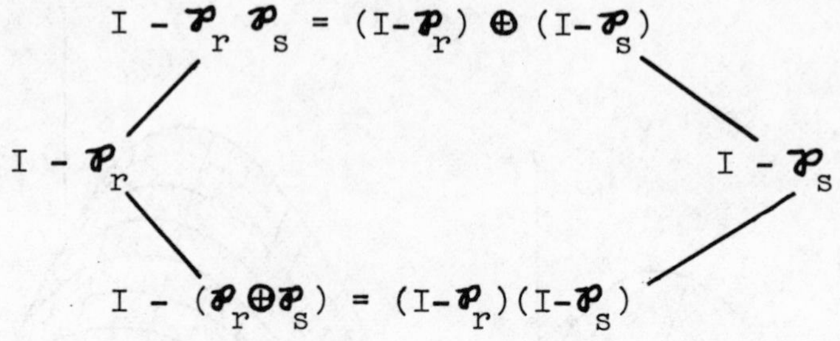

Figure 7. Lattice of remainders.

The following very general results of Cavendish, Gordon, and Hall [3] for L-splines establish the validity of this inference. (For cubic spline blending see also [2] and for error bounds based on generalizations of the Bramble-Hilbert Lemma, see Watkins [15]).

L-splines have been given considerable attention in the approximation theory literature [9-14]. Let

$$\pi_x: \quad 0 = x_0 < x_1 < \ldots < x_N = 1 \tag{10}$$

be a partition of [0,1] with mesh gauge and mesh ratio

$$\begin{aligned} h_x &\equiv \max_i |x_i - x_{i-1}|, \\ \beta_x &\equiv h_x / \min_i |x_i - x_{i-1}| \end{aligned} \tag{11}$$

respectively. Consider the nth order differential operator L_x defined by

$$L_x[u](x) \equiv \sum_{j=0}^{n} c_j(x)\, D^j u(x), \; n \geq 1 \tag{12}$$

where $c_j(x) \in C^j[0,1]$, $0 \leq j \leq n$ and $c_n(x) \geq \delta > 0$ for $x \in [0,1]$. The <u>L-spline</u> <u>space</u>, $S(L_x, \pi_x, z_x)$ associated with the operator L_x, partition π_x and integer $z_x (1 \leq z_x \leq n)$, is defined to be the collection of all real-valued functions $w(x)$ defined on [0,1] such that

$$L_x^* L_x[w](x) = 0, \quad x \in (0,1) - \{x_i\}_{i=1}^{N-1}$$

$$D^k w(x_i -) = D^k w(x_i +),$$
$$0 \leq k \leq 2n - 1 - z_x, \quad 0 < i < N$$

where $L_x^*[v](x) \equiv \sum_{j=0}^{n} (-1)^j D^j\{c_j(x)v(x)\}$ is the formal adjoint of L_x.

From Varga [14] and Swartz and Varga [13] we have $S(L_x,\pi_x z_x) \subset W_\infty^{2n-z_x}[0,1]$, dim $S(L_x,\pi_x,z_x)=2n+z_x(N-1)$ and given $g \in W_2^{2n}[0,1]$, there exists a unique element $s \in S(L_x,\pi_x,z_x)$ such that

$$\text{(i)} \quad D^j[g-s](x_i)=0,\ 0\le j\le z_x-1,\ 0\le i<N$$

$$\text{(ii)} \quad D^j[g-s](0)=D^j[g-s](1)=0,\ 0\le j\le n-1 \qquad (13)$$

$$\text{(iii)} \quad ||D^j[g-s]||_{L_2[0,1]} \le K_x h_x^{2n-j}||g||_{W_2^{2n}[0,1]},$$

$$0\le j\le 2n-1,$$

where K_x is a constant and β_x is bounded as $h_x \to 0$. Choosing for example, $L_x = D^2$, $z_x = 1$ yields the cubic spline space. These univariate convergence results are used to establish the following analogous results for the transfinite Boolean sum interpolant.

THEOREM (Cavendish, Gordon, Hall [3]): Let $\mathcal{P}_r$ and $\mathcal{P}_s$ be commuting projectors from $W_2^{(2n,2m)}(\mathcal{S})$ onto $S(L_r,\pi_r,z_r)$ and $S(L_s,\pi_s,z_s)$ respectively, determined by the interpolation schemes (13i,13ii). Then the Boolean sum projection $s_f \equiv (\mathcal{P}_r \oplus \mathcal{P}_s)[f]$ satisfies the following interpolation properties:

$$(f-s_f)^{(k,0)}(r_i,s) = 0,$$

$$0 \le k \le z_r - 1,$$

$$0 < i < N,\quad s \in [0,1],$$

$$\text{(i)} \quad (f-s_f)^{(0,\ell)}(r,s_j) = 0,$$

$$0 \le \ell \le z_s - 1,$$

$0 < j < M, \quad r \in [0,1]$,

$$(f-s_f)^{(k,0)}(0,s) \qquad (14)$$

$$= (f-s_f)^{(k,0)}(1,s) = 0,$$

$0 \le k \le m-1, \quad s \in [0,1]$,

(ii) $$(f-s_f)^{(0,\ell)}(r,0)$$

$$= (f-s_f)^{(0,\ell)}(r,1) = 0,$$

$0 \le \ell \le n-1, \; r \in [0,1]$,

<u>where</u> $f \in W_2^{(2n,2m)}(\mathcal{S})$. <u>Moreover</u>, <u>if</u> β_r <u>and</u> β_s <u>are</u> <u>bounded</u> <u>as</u> $h \to 0$, <u>then</u>

$$||(f-\mathcal{P}_r \oplus \mathcal{P}_s[f])^{(k,\ell)}||_{L_2(\mathcal{S})}$$

$$\le K_r K_s h_r^{2n-k} h_s^{2m-\ell} ||f||_{W_2^{(2n,2m)}(\mathcal{S})} \qquad (15)$$

$$= O(h_r^{2n-k} h_s^{2m-\ell})$$

<u>for</u> $0 \le k \le 2n-1$ <u>and</u> $0 \le \ell \le 2m-1$.

Note that (15) is consistent with the algebraic result given in Figure 7 (which applies in this more general L-spline setting as well) that the remainder $I - (\mathcal{P}_r \oplus \mathcal{P}_s)$ is the product of remainders $(I-\mathcal{P}_r)$ and $(I-\mathcal{P}_s)$.

As in [5,6,8] the functions $f(r_i,s)$, $0 \le i \le N$ and $f(r,s_j)$ $0 \le j \le M$ can now be approximated by finite parameter interpolation as follows.

Let $\bar{\pi}_r$: $0 = \bar{r}_0 < \bar{r}_1 < \ldots < \bar{r}_{\bar{N}} = 1$ and $\bar{\pi}_s$: $0 = \bar{s}_0 < \bar{s}_1 < \ldots < \bar{s}_{\bar{M}} = 1$ be partitions of $[0,1]$ of mesh gauges $\bar{h}_r$ and $\bar{h}_s$ respectively, and such that $\bar{\pi}_r$ is a <u>refinement</u> of π_r and $\bar{\pi}_s$ is a refinement of π_s. Choose operators $\bar{L}_r$ and $\bar{L}_s$ and integers $\bar{z}_r \ge z_r$, $\bar{z}_s \ge z_s$. If $\bar{\mathcal{P}}_r, \bar{\mathcal{P}}_s$ are the corresponding L-spline interpolation projectors, then the discrete $(L_r, z_r; L_s, z_s; \bar{L}_r, \bar{z}_r; \bar{L}_s, \bar{z}_s)$ blended interpolant S_f to f, is given by

$$S_f \equiv \left(\bar{\mathcal{P}}_s \mathcal{P}_r + \bar{\mathcal{P}}_r \mathcal{P}_s - \mathcal{P}_r \mathcal{P}_s\right)[f], \tag{16}$$

and we have

THEOREM 2 (Cavendish, Gordon, Hall [3]): The discrete $(L_r, z_r; L_s, z_s; \bar{L}_r, \bar{z}_r; \bar{L}_s, \bar{z}_s)$ blended interpolant S_f defined in (16) interpolates to f in the sense that

$$\begin{aligned} &(S_f - f)^{(k,\ell)}(r_i, \bar{s}_j) = 0 \\ &0 \le k \le z_r - 1,\ 0 < i < N \\ &0 \le k \le n-1,\ i = 0, N \\ &0 \le \ell \le \bar{z}_s - 1,\ 0 < j < \bar{M} \\ &0 \le \ell \le \bar{m}-1,\ j = 0, \bar{M} \end{aligned} \tag{17}$$

and

$$\begin{aligned} &(S_f - f)^{(k,\ell)}(\bar{r}_i, s_j) = 0 \\ &0 \le k \le \bar{z}_r - 1,\ 0 < i < \bar{N} \\ &0 \le k \le \bar{n}-1,\ i = 0, \bar{N} \\ &0 \le \ell \le z_s - 1,\ 0 < j < M \\ &0 \le \ell \le m-1,\ j = 0, M. \end{aligned} \tag{18}$$

Further, if the projectors $\mathcal{P}_r, \mathcal{P}_s, \bar{\mathcal{P}}_r, \bar{\mathcal{P}}_s$ satisfy $\mathcal{P}_r\mathcal{P}_s = \mathcal{P}_s\mathcal{P}_r$, $\bar{\mathcal{P}}_r\mathcal{P}_s = \mathcal{P}_s\bar{\mathcal{P}}_r$, and $\mathcal{P}_r\bar{\mathcal{P}}_s = \bar{\mathcal{P}}_s\mathcal{P}_r$ and if $f \in W_2^{(2p,2q)}(\mathcal{S})$, where $p = \max(n,\bar{n})$ and $q = \max(m,\bar{m})$, then

$$||(f - S_f)^{(k,\ell)}||_{L_2(\mathcal{S})} = O(\bar{h}_r^{2\bar{n}-k} + \bar{h}_s^{2\bar{m}-\ell} + h_r^{2n-k} h_s^{2m-\ell})$$

as $h \equiv \max(h_r, h_s) \to 0$, $0 \le k \le 2\min(n,\bar{n}) - 1$ and $0 \le \ell \le 2\min(m,\bar{m}) - 1$.

Proof: As in Gordon [6, Lemma A1],

$$f-S_f = \left(I-(\bar{\mathcal{P}}_s\mathcal{P}_r+\bar{\mathcal{P}}_r\mathcal{P}_s-\mathcal{P}_r\mathcal{P}_s)\right)[f]$$
$$= \Big((I-\bar{\mathcal{P}}_r)+(I-\bar{\mathcal{P}}_s)+(I-\mathcal{P}_r\oplus\mathcal{P}_s)$$
$$- (I-\bar{\mathcal{P}}_r\oplus\mathcal{P}_s)-(I-\mathcal{P}_r\oplus\bar{\mathcal{P}}_s)\Big)[f].$$

The desired result follows from (15) and (13iii). Q.E.D.

If we choose $\bar{\mathcal{P}}_r = \mathcal{P}_r$ and $\bar{\mathcal{P}}_s = \mathcal{P}_s$, then $S_f \equiv \mathcal{P}_r\mathcal{P}_s$, the tensor product projector. In this case (19) yields

$$||(f-\mathcal{P}_r\mathcal{P}_s)^{(k,\ell)}||_{L_2(\mathcal{S})} = O(h_r^{2n-k}+h_s^{2m-\ell}+h_r^{2n-k}h_s^{2m-\ell})$$

This is consistent with the algebraic result in Figure 7 that the remainder $I - \mathcal{P}_r\mathcal{P}_s = (I-\mathcal{P}_r) \oplus (I-\mathcal{P}_s)$.

For the spline blended surface $\mathcal{P}_r \oplus \mathcal{P}_s[f]$, the operators $\bar{L}_r$ and $\bar{L}_s$ are typically chosen again as D^2, however, $\bar{\pi}_r$ (resp. $\bar{\pi}_s$) is a much finer mesh than π_r (resp. π_s). This amounts to approximating the functions $f(r_i,s)$, $0 \leq i \leq N$ and $f(r,s_j)$, $0 \leq j \leq M$ by cubic splines on a very fine partition. Such a spline blended network of cubic splines is called a "Gordon Surface" after W. J. Gordon who developed programs for Fisher Body-GM based on such interpolation schemes in the middle to late '60's.

In fact, work in the area of bivariate interpolation on surface representation was spearheaded by the Mathematics Department of General Motors Research Laboratories over the past 15 years, [1,4,5]. Automobile exterior surfaces are extremely "smooth" and an explicit mathematical representation to be used in computer-aided design necessitates constructable surfaces of interpolation with continuous curvature as well as continuous gradient.

Typically, the given data consists of a set of points

$$\Big((r_i,s_j,f(r_i,s_j)):\ 0 \leq i \leq p_1,\ 0 \leq j \leq p_2\Big)$$

and due to their regularity, the tensor product scheme (3) can easily be implemented with $\phi_i(r)$ and $\psi_j(s)$ being cubic splines, $N = p_1$ and $M = p_2$. The resulting surface, passes through the given data, is of continuity class C^2 and is termed the bicubic spline interpolant of the given data, [1,4]. Alternatively, we can choose $N \ll p_1$, $M \ll p_2$, construct a network of cubic splines $\{f(r_i,s)\}_{i=0}^{N}$ $\{f(r,s_j)\}_{j=0}^{M}$ and blend these together using cubic spline blending functions $\phi_i(r)$ and $\psi_j(s)$. The resulting cubic spline blended [5] or Gordon surface is also of continuity class C^2, but in general its construction is dependent on far fewer data than the bicubic spline. The stencils given in Figure 8 are indicative of the data used to construct the two surfaces and in

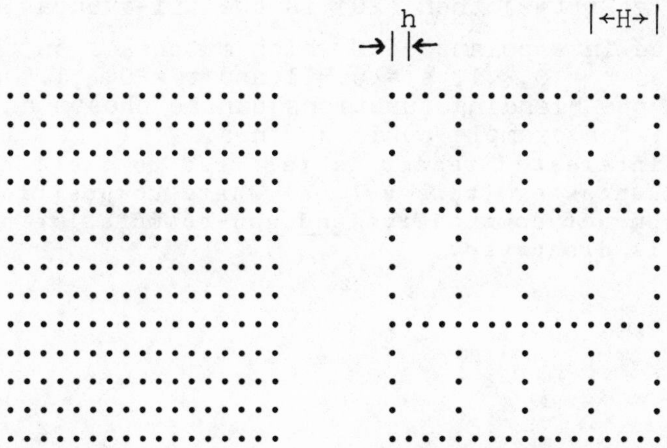

Figure 8. Data at 221 points determine the bicubic spline surface while data at 113 points determine the cubic spline blended surface of comparable accuracy.

this connection the Boolean sum interpolant is transfinite in the sense that the network of cubic splines are matched exactly. Theorem 2 proves if we choose $h = H^2$ in Figure 8 then asymptotically (as $h \to 0$) the bicubic spline $\mathcal{P}_r\mathcal{P}_s[f]$ and spline blended interpolant $\mathcal{P}_r \oplus \mathcal{P}_s[f]$ both converge to a function $f \in C^{(4,4)}$ at the same rate, $O(h^4)$.

If f is a function of 3 variables then the transfinite interpolation formula analogous to (8) is

$$(\mathcal{P}_r \oplus \mathcal{P}_s \oplus \mathcal{P}_t)[f] \equiv \mathcal{P}_r[f] + \mathcal{P}_s[f] + \mathcal{P}_t[f] - \mathcal{P}_r\mathcal{P}_t[f] - \mathcal{P}_s\mathcal{P}_t[f] - \mathcal{P}_r\mathcal{P}_s[f] + \mathcal{P}_r\mathcal{P}_s\mathcal{P}_t[f] \qquad (20)$$

$$\mathcal{P}_r[f] = \sum_{i=0}^{M} \phi_i(r)\, f(r_i,s,t)$$

$$\mathcal{P}_s[f] = \sum_{j=0}^{N} \psi_j(s)\, f(r,s_j,t)$$

$$\mathcal{P}_t[f] = \sum_{k=0}^{P} \eta_k(t)\, f(r,s,t_k)$$

For example, if $\phi_i(r)$, $\psi_j(s)$ and $\eta_k(t)$ are quadratic polynomials, e.g. $\eta_0(t) = 2(t-\frac{1}{2})(t-1)$, $\eta_1(t) = 4t(1-t)$ and $\eta_2(t) = 2t(t-\frac{1}{2})$ then (20) is the tri-quadratic transfinite interpolant of f which matches f on the nine planes $r = 0,\frac{1}{2},1$, $s = 0,\frac{1}{2},1$ and $t = 0,\frac{1}{2},1$. Alternatively, the blending functions can be chosen as L-splines, for example cubic splines.

The interested reader is referred to the lectures of Peter Lancaster (this volume) where composition and Boolean sums of commutative <u>and</u> non-commutative projectors is discussed.

REFERENCES

1. G. Birkhoff and C. DeBoor, "Piecewise Polynomial Interpolation and Approximation," in Approximation Functions, H. Garabedian, ed., 164-190 Amsterdam, Elsevier, 1965.

2. R. E. Carlson and C. A. Hall, "Error Bounds for Bicubic Spline Interpolation," J. Approx. Theory, 7, 41-47, (1973).

3. J. C. Cavendish, W. J. Gordon, and C. A. Hall, "Ritz-Galerkin Approximations in Blending Function Spaces," Numer. Math., 26, 155-178 (1976).

4. C. DeBoor, "Bicubic Spline Interpolation," J. Math. and Phys., 41, 212-218, (1962).

5. W. J. Gordon, "Spline-Blended Surface Interpolation Through Curve Networks," J. Math. Mech., 10, 931-952, (1968).

6. W. J. Gordon, "Distributive Lattices and the Approximation of Multivariate Functions," in Approximation with Special Emphasis on Spline Functions, 223-277, I. J. Schoenberg, ed., New York, Academic Press, 1969.

7. W. J. Gordon and C. A. Hall, "Transfinite Element Methods: Blending-Function Interpolation Over Curved Element Domains," Numer. Math. 21, 109-129, (1973).

8. C. A. Hall, "Transfinite Interpolation and Applications to Engineering Applications," in Theory of Approximation with Applications, eds., A. Law and B. Sahney, Academic Press, 1976.

9. M. Schultz and R. S. Varga, "L-splines," Numer. Math., 10, 345-369, (1967).

10. M. Schultz, "L^∞-multivariate Approximation Theory," SIAM J. Numer. Analysis, 6, 161-183, (1969).

11. M. Schultz, "L^2-multivariate Approximation Theory," SIAM J. Numer. Analysis, 6, 184-209, (1969).

12. M. Schultz, Spline Analysis, Prentice-Hall, Inc., Englewood Cliffs, New Jersey, 1973.

13. B. K. Swartz and R. S. Varga, "Error Bounds for Spline and L-spline Interpolation," J. Approx. Theory 6, 6-149, (1972).

14. R. S. Varga, Functional Analysis and Approximation Theory in Numerical Analysis, SIAM, 1972.

15. D. S. Watkins, "Error Bounds for Polynomial Blending Function Methods," SIAM J. Numer. Analysis, 14, 721-734.

APPENDIX: PROGRAM BLEND

Charles Hall and Frank Sledge

BLEND is a FORTRANIV program which plots one of a variety of blended surfaces $\mathcal{P}_r \oplus \mathcal{P}_s[f]$ passing through a given network of curves

$$\text{FISAI} \equiv f(r_i, s) \quad i = 1, \ldots, \text{NRAC}$$

$$\text{GJRAI} \equiv f(r, s_j) \quad j = 1, \ldots, \text{NSAC}.$$

The program utilizes a Tektronix Graphics Terminal and requires the PLOT-10 Software System as well as routines ICSICU, ICSEVU and IERTST from the IMSL software package.

The surface plotted is (Cf. (8))

$$\mathcal{P}_r \oplus \mathcal{P}_s[f] = \sum_{i=1}^{\text{NRAC}} f(r_i,s)p_i(r) + \sum_{j=1}^{\text{NSAC}} f(r,s_j)q_j(s) - \sum_{i=1}^{\text{NRAC}} \sum_{j=1}^{\text{NSAC}} f(r_i,s_j)p_i(r)q_j(s)$$

There are two groups of input to BLEND:

I. User will be prompted for unit number of control file ICTLFL ($10 \leq \text{ICTLFL} \leq 63$) consisting of 21 numbers which determine the surface.

IPTRC = 1 Trace all surface points in r,s,t coordinates onto a trace file (unit 9)

= 0 No trace wanted

GAMMA = angle of display, rotation about t-axis $(-360 \leq \gamma \leq 360)$.

THETA = angle of display, forward tilt of t-axis $(-90 \leq \theta \leq 90)$

NRAC = Number of given r curves (displayed as solid lines)

NRAI = Number of r curves between given $f(r_i,s)$ curves, (displayed as dotted lines)

NSAP = Number of s points where r curves are to be evaluated

NSAC = Number of given s curves (displayed as solid lines)

NSAI = Number of s curves between given $f(r,s_j)$ curves, (displayed as dotted lines)

NRAP = Number of r points where s curves are to be evaluated

II. Each class of functions $f(r_i,s)$, $f(r,s_j)$, $p_i(r)$ and $q_j(s)$ require three parameters to be specified. These are stored internally in the matrix IDATFL (3,4)

Index K = 1 corresponds to the r curves $f(r_i,s)$

K = 2 corresponds to the s curves $f(r,s_j)$

K = 3 corresponds to the blending functions $p_i(r)$

K = 4 corresponds to the blending functions $q_j(s)$

IDATFL(1,K): abscissae of r(K=1) or s(K=2) curves.

= 0 evenly spaced on [0,1]

= 1 Read from unit number IDATFL(3,K)

= 2 evenly spaced on [a,b], a,b given on unit number IDATFL(3,K)

IDATFL(2,K): function definition

= 0 zero function

= 1 user supplied subroutine EVFCTN
piecewise linear spline of unit height at specific abscissae

= 3 polynomial of degree $\leq$ 99. Coefficients and degree on unit IDAFL(3,K)

= 4 natural cubic spline of unit height at specific abscissae

= 5 cubic spline with 100 or less knots. Data on unit IDATFL(3,K)

IDATFL(3,K): Unit number of data file, if needed.

The surface will be displayed on a "support block" showing coordinate axes. The entire display will be automatically scaled, centered and rotated as needed. A hidden line removal algorithm will be applied to the support block but not the surface itself.

The user supplied subroutine is EVFCTN(K,Y,U,W,M,N) where $1 \leq K \leq 4$ as above, V(N) are the abscissae associated with given curves, U(M) are the points where function is to be evaluated and W(M,N) contains the values of the N functions at the M points.

A listing of this program can be obtained from the authors.

VECTOR-VALUED POLYNOMIAL AND SPLINE APPROXIMATION

Charles Hall*
Institute for Computational Mathematics
and Applications
Department of Mathematics and Statistics
University of Pittsburgh

ABSTRACT Blended interpolation of vector-valued functions is discussed relative to their applications in the construction of finite elements, mesh generation and table look-up of thermodynamic properties of a gas or liquid. The invertibility of two isoparametric transformations is investigated.

INT. Much of the following material is taken from joint papers with W.J. Gordon [9,10] and A.E. Frey and T.A. Porsching [4].

To begin, let us recall the geometric interpretation of the graph of a vector-valued function of two independent variables r and s

$$\underline{F}(r,s) = [x_1(r,s), x_2(r,s), \ldots, x_n(r,s)]^T. \qquad (1)$$

As the variables r and s range over a domain $\mathcal{S}$ in the r,s-plane R^2, $\underline{F}(r,s)$ traces out a region $\mathcal{R}$ in Euclidean n-space, E^n. That is, $\underline{F}$ maps regions in R^2 into regions in E^n.

$$\underline{F}: R^2 \to E^n. \qquad (2)$$

For the most part, we shall be concerned with continuous transformations $\underline{F}$ which map the unit square $\mathcal{S} = [0,1] \times [0,1]$ one-to-one onto a simply connected, bounded region $\mathcal{R}$ in E^2 or E^3. Such maps can be thought of as topological distortions of the planar region $\mathcal{S}$ onto the two-dimensional manifold $\mathcal{R}$, which is either a planar region ($\mathcal{R} \subset E^2$) or a surface embedded in 3-space

*Partially supported by USAF Office of Scientific Research Contract No. F44620-76-C-0104.

Badri N. Sahney (ed.), Polynomial and Spline Approximation, 35-67.

($\mathcal{R} \subset E^3$). In either case a one-to-one (univalent) mapping $\mathcal{S} \to \mathcal{R}$ is equivalent to the introduction of a curvilinear co-ordinate system on $\mathcal{R}$. The curve of constant generalized co-ordinate $r = r^*$ is the image $\underline{F}(r^*,s)$ of the line $r = r^*$ in $\mathcal{S}$.

Similarly, the curve $\underline{F}(r,s^*)$ is the set of all points in $\mathcal{R}$ with generalized co-ordinate $s = s^*$. Thus the point $\underline{F}(r^*,s^*)$ on $\mathcal{R}$ is said to have generalized co-ordinates (r^*,s^*), and, since the mapping $\mathcal{S} \to \mathcal{R}$ is univalent, any point $P \in \mathcal{R}$ can be uniquely referenced by its generalized co-ordinates.

If $\mathcal{S}$ is the unit cube $[0,1] \times [0,1] \times [0,1]$ in the r,s,t-parameter space R^3 and $\mathcal{R}$ is a bounded region in Euclidean 3-space, then a one-to-one mapping $\underline{F}$ of $\mathcal{S}$ onto $\mathcal{R}$ can be envisioned as a topological distortion of the cube into $\mathcal{R}$. Such a mapping $R^3 \to E^3$ generates a curvilinear co-ordinatization of the solid $\mathcal{R}$ so that each point of $\mathcal{R}$ may be referenced by its generalized co-ordinates (r,s,t).

By a canonical region in Euclidean 2- or 3-space, we mean a region for which there exists a classical curvilinear co-ordinate system such as Cartesian, polar, bipolar or elliptic in E^2 and such as Cartesian, cylindrical, spheroidal or torodial in E^3 [12, p. 655]. The very familiarity of these co-ordinate systems leads one to lose sight of the basic fact that they are simply one-to-one mappings of a rectangle or rectangular parallelepiped (in the parameter domain) onto the canonical region in Euclidean space. Consider, for instance, the mapping

$$\underline{F}(r,s) = \begin{bmatrix} r \cos 2\pi s \\ r \sin 2\pi s \end{bmatrix}, \quad r \varepsilon (0,1], \; s \varepsilon (0,1] \qquad (3)$$

which maps the unit square $\mathcal{S}$: $(0,1] \times (0,1]$ in R^2 one-to-one onto the unit disc in Euclidean 2-space and thereby induces a polar co-ordinatization.

If the problem domain $\mathcal{R}$ is geometrically reminiscent of some canonical region, then it is reasonable to seek a co-ordinatization of $\mathcal{R}$ which is, in some sence, a mild distortion of the natural co-ordinate system of the canonical domain. The plane region in Figure 1(a), for example,

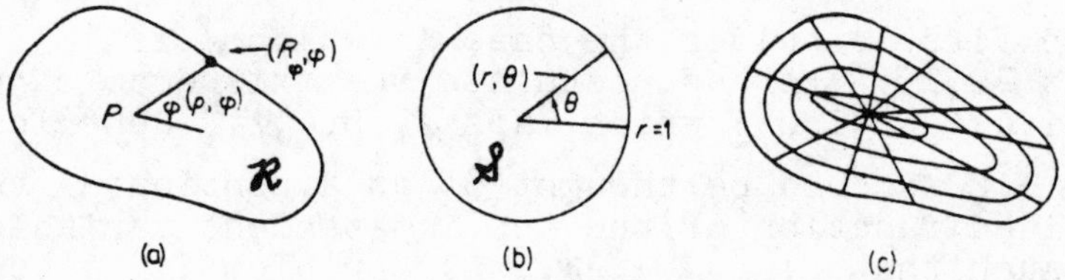

Figure 1. (a) A non-canonical region $\mathcal{R}$ in E^2. (b) The unit disc $\mathcal{S}$ in R^2. (c) The curvilinear co-ordinate system induced on $\mathcal{R}$ by (4).

can be regarded as a distorted circle. Since the region is star-shaped with respect to the indicated point P, the simple radially expansive map $\underline{T}$: $\mathcal{S} \to \mathcal{R}$ given by

$$\phi \equiv \theta, \quad \rho \equiv rR_\phi \tag{4}$$

puts the points of $\mathcal{R}$ into one-to-one correspondence with the points of the unit disc $\mathcal{S}$. The images in $\mathcal{R}$ of the constant co-ordinate lines

$$r = \text{constant}, \quad \theta = \text{constant} \tag{5}$$

are 'concentric contours' and radial lines as shown in Figure 1(c). Another perspective on the same problem is that any point in $\mathcal{R}$ may now be referenced by specification of its generalized polar co-ordinates (ρ,ϕ). The map $\underline{T}$: $\mathcal{S} \to \mathcal{R}$ has defined a curvilinear co-ordinatization of $\mathcal{R}$.

For bounded, simply connected domains $\mathcal{R}$, one could of course generate an orthogonal co-ordinatization by means of a conformal mapping of $\mathcal{R}$ onto a canonical region such as a circle or a square. However, from a practical point of view, the construction of a conformal map is in fact equivalent to the solution of Laplace's equation and is contrary to our overall goals of computational simplicity. Thus, we shall be concerned herein with mapping techniques which are relatively simple to construct and implement for a wide variety of regions. These mappings are based upon the multivariate interpolation techniques originally developed for the representation and approximation of complex geometric shapes in computer-aided design and numerically controlled machining applications; see [8] and the references cited therein.

We first consider the case in which $\mathcal{S}$ is $[0,1] \times [0,1]$. Let us postulate the existence of a primitive function $\underline{F}$ which maps $\mathcal{S}$ onto $\mathcal{R}$. Generically, $\underline{F}$: $R^2 \to E^2$ should be thought of as a continuous vector-valued function of the two independent variables r and s such that $\underline{F}$: $\partial\mathcal{S} \to \partial\mathcal{R}$. Our problem is to construct a univalent (one-to-one) function $\underline{U}$: $\mathcal{S} \to \mathcal{R}$ which matches $\underline{F}$ on the boundary of $\mathcal{S}$, i.e.

$$\begin{aligned} \underline{U}(0,s) &= \underline{F}(0,s), \quad \underline{U}(r,0) = \underline{F}(r,0) \\ \underline{U}(1,s) &= \underline{F}(1,s), \quad \underline{U}(r,1) = \underline{F}(r,1) \end{aligned} \tag{6}$$

A function $\underline{U}$ which interpolates to $\underline{F}$ at a non-denumerable set of points is termed a transfinite interpolant of $\underline{F}$.

Interpolation problem (6) can be viewed as a search for a projector $\mathcal{P}$ such that $\underline{U} = \mathcal{P}[F]$ is a univalent map of $\mathcal{S} \to \mathcal{R}$ which satisfies the desired interpolatory properties. $\underline{U}$ is termed the projection of $\underline{F}$ or the image of $\underline{F}$ under $\mathcal{P}$.

Suppose now that ϕ_0, ϕ_1 and ψ_0, ψ_1 are four univariate functions which satisfy the cardinality conditions

$$\begin{aligned} \phi_i(r_k) &= \delta_{ik} \equiv \begin{cases} 1, & i = k \\ 0, & i \neq k \end{cases} & \text{for } i,k = 0,1 \\ \psi_j(s_\ell) &= \delta_{j\ell} & \text{for } j,\ell = 0,1 \end{aligned} \tag{7}$$

where, unless otherwise stated, $r_0 = s_0 = 0$ and $r_1 = s_1 = 1$; and consider the projectors $\mathcal{P}_r$ and $\mathcal{P}_s$ defined by

$$\begin{aligned} \mathcal{P}_r[\underline{F}] &\equiv \phi_0(r)\underline{F}(r_0,s) + \phi_1(r)\underline{F}(r_1,s) \\ \mathcal{P}_s[\underline{F}] &\equiv \psi_0(s)\underline{F}(r,s_0) + \psi_1(s)\underline{F}(r,s_1) \end{aligned} \tag{8}$$

Then, the product projection

$$\mathcal{P}_r\mathcal{P}_s[\underline{F}] = \sum_{i=0}^{1} \sum_{j=0}^{1} \underline{F}(r_i,s_j)\phi_i(r)\psi_j(s) \tag{9}$$

interpolates to $\underline{F}$ at the four corners of $[r_0,r_1] \times [s_0,s_1]$ and the Boolean sum projection

$$(\mathcal{P}_r \oplus \mathcal{P}_s)[\underline{F}] \equiv \mathcal{P}_r[\underline{F}] + \mathcal{P}_s[\underline{F}] - \mathcal{P}_r \mathcal{P}_s[\underline{F}] \tag{10}$$

interpolates to $\underline{F}$ on the entire boundary of $[r_0,r_1] \times [s_0,s_1]$. These properties of the functions (9) and (10) may be readily verified by evaluating the right-hand sides for the appropriate values of r and s and recalling the cardinality properties (7); see also [6,7,8].

Since the function $\underline{U} = (\mathcal{P}_r \oplus \mathcal{P}_s)[\underline{F}]$ matches $\underline{F}$ on the non-denumerable set of points comprising the boundary of $\mathcal{S} = [r_0,r_1] \times [s_0,s_1]$, $\underline{U}$ is a transfinite interpolant of $\underline{F}$. Note that, in sharp contrast, the projection $\mathcal{P}_r \mathcal{P}_s[\underline{F}]$ matches $\underline{F}$ only at a finite number of distinct points - namely, the four corners of the region.

The functions ϕ_i and ψ_j in the above formulae are as yet unspecified except for their values at the points $r_0 = s_0 = 0$ and $r_1 = s_1 = 1$. They are commonly referred to as 'blending functions' and the function $\underline{U} = (\mathcal{P}_r \oplus \mathcal{P}_s)[\underline{F}]$ is termed a blending interpolant. The simplest choice for the blending functions in (7) is (as in the scalar case)

$$\begin{aligned} \phi_0(r) &= 1 - r, \quad \psi_0(s) = 1 - s \\ \phi_1(r) &= r, \qquad \psi_1(s) = s \end{aligned} \tag{11}$$

The vector-valued bivariate function $\underline{U} \equiv \mathcal{P}_r \oplus \mathcal{P}_s[\underline{F}]$ obtained by using (10) and (11) is termed the <u>bilinearly blended interpolant</u> of $\underline{F}$, or the <u>transfinite bilinear Lagrange interpolant</u> of $\underline{F}$. Explicitly, it is given by

$$\begin{aligned} \underline{U}(r,s) = {} & (1-r)\underline{F}(0,s) + r\underline{F}(1,s) + (1-s)\underline{F}(r,0) + s\underline{F}(r,1) \\ & -(1-r)(1-s)\underline{F}(0,0)-(1-r)s\underline{F}(0,1)-r(1-s)\underline{F}(1,0)-rs\underline{F}(1,1). \end{aligned} \tag{12}$$

The reader who is familiar with the work of Coons will recognize this as the simplest of the 'Coons patches' described first in [2]. It has the property that $\underline{U} = \underline{F}$ on the perimeter of the unit square $[0,1] \times [0,1]$.

Isoparametric Elements.

Formula (12) coupled with the scalar equation (13) below are useful in constructing and analyzing finite elements with curved boundaries. Cf. Figure 2.

We recall, cf. [10], that choosing linear or quadratic blending functions in equations (4) and (5) of lecture 1 the scalar blending function theory [(8) of lecture 1] produces the finite element stencils in Table 2 for appropriately chosen polynomials $f(r_i, s)$ and $f(r, s_j)$. Details of such constructions along with error analyses are given in [10]. See also [14].

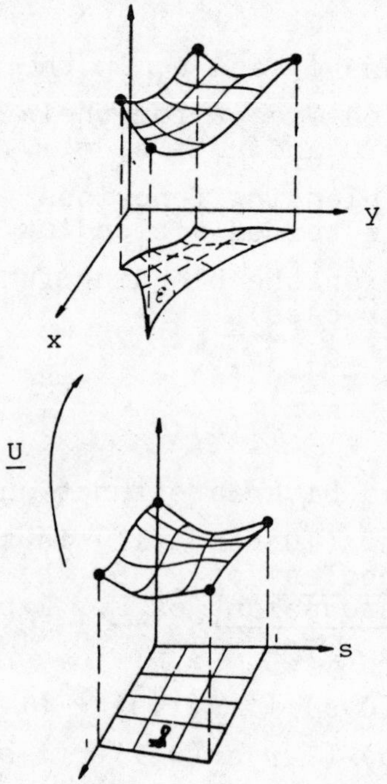

Figure 2. A univalent mapping U induces a curvilinear coordinate system on the domain E and the interpolation-approximation problem can be described and then solved in terms of the r, s system.

Boundary Curves	Finite Element Stencil	Asymptotic Order	Curves $f(r_i,s)$ $f(r,s_j)$	Finite Element Stencil	Asymptotic Order
linear		2	quadratic	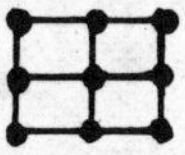	3
quadratic	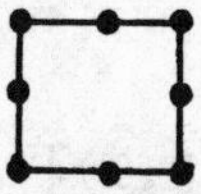	3	quartic	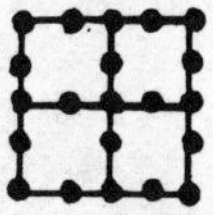	5
cubic	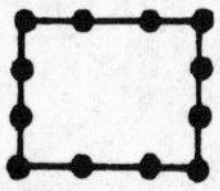	4	quintic	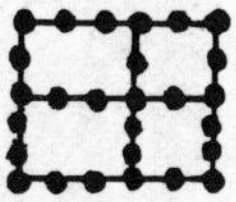	6
quartic		4			
Finite element stencils - bilinear blending.			Finite element stencils - biquadratic blending.		

$$\mathcal{P}_r \oplus \mathcal{P}_s[f] = \sum_{i=0}^{N} f(r_i,s)\phi_i(r) + \sum_{j=0}^{M} f(r,s_j)\psi_j(s) - \sum_{i=0}^{N}\sum_{j=0}^{M} f(r_i,s_j)\phi_i(r)\psi_j(s). \tag{13}$$

Bilinear blending:

$$\phi_0(r) = 1 - r, \quad \phi_1(r) = r,$$
$$\psi_0(s) = 1 - s, \quad \psi_1(s) = s.$$

Biquadratic blending:

$$\phi_0(r) = 2(\tfrac{1}{2}-r)(1-r), \quad \phi_1(r) = 4r(1-r),$$
$$\phi_2(r) = 2r(r-\tfrac{1}{2}), \quad \psi_j(s) = \phi_j(s), \; j = 0,1,2.$$

Table 2. Finite elements generated by blending a network of curves.

If $\mathcal{E}$ is the domain in Figure 3 with boundary segments parametrized as indicated, then define $\underline{U}$: $\mathcal{S} \to \mathcal{E}$, where

$$\underline{U}; \quad \begin{bmatrix} x(r,s) \\ y(r,s) \end{bmatrix} = (1-r) \begin{bmatrix} x(0,s) \\ y(0,s) \end{bmatrix} + r \begin{bmatrix} x(1,s) \\ y(1,s) \end{bmatrix}$$

$$+ (1-s) \begin{bmatrix} x(r,0) \\ y(r,0) \end{bmatrix} + s \begin{bmatrix} x(r,1) \\ y(r,1) \end{bmatrix} \tag{14}$$

$$- (1-r)(1-s) \begin{bmatrix} x(0,0) \\ y(0,0) \end{bmatrix} - (1-r)s \begin{bmatrix} x(0,1) \\ y(0,1) \end{bmatrix}$$

$$- r(1-s) \begin{bmatrix} x(1,0) \\ y(1,0) \end{bmatrix} - rs \begin{bmatrix} x(1,1) \\ y(1,1) \end{bmatrix}.$$

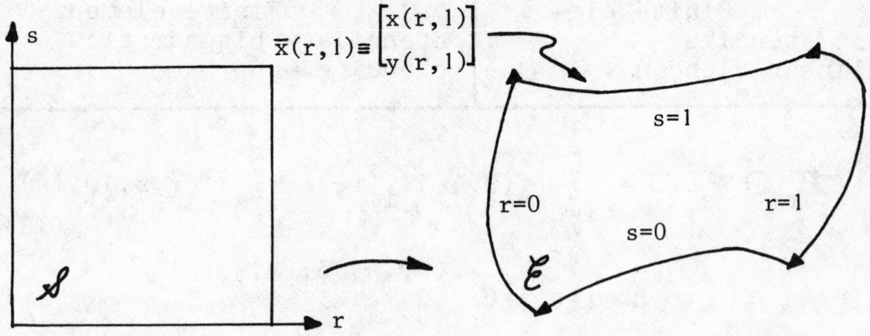

Figure 3. Domain Transformation

If the boundary curves, e.g. $\bar{x}(0,s) = \begin{bmatrix} x(0,s) \\ y(0,s) \end{bmatrix}$, are parametrized as polynomials of the same form as used to approximate the dependent variable f (by, say u) and also the same blending functions are used then we obtain the isoparametric formulae of Zienkiewicz [16], (cf. Figure 4). Similarly, sub- and super-parametric elements can be

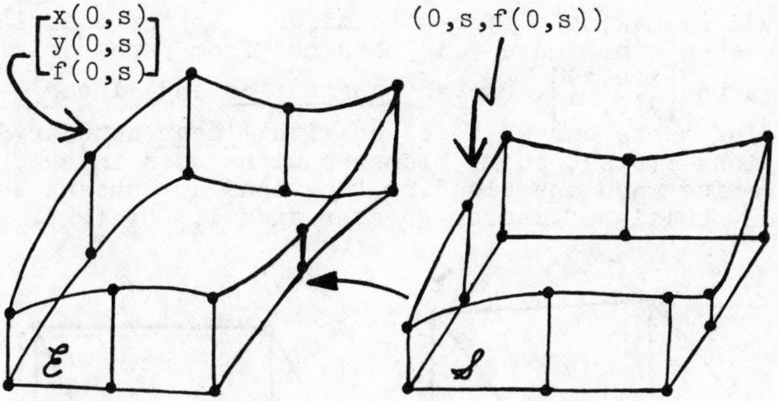

Figure 4. An isoparametric element generated from the transfinite formula:

$$\begin{bmatrix} x(r,s) \\ y(r,s) \\ u(r,s) \end{bmatrix} = (1-r) \begin{bmatrix} x(0,s) \\ y(0,s) \\ f(0,s) \end{bmatrix} + r \begin{bmatrix} x(1,s) \\ y(1,s) \\ f(1,s) \end{bmatrix}$$

$$+ (1-s) \begin{bmatrix} x(r,0) \\ y(r,0) \\ f(r,0) \end{bmatrix} + s \begin{bmatrix} x(r,1) \\ y(r,1) \\ f(r,1) \end{bmatrix}$$

$$- (1-r)(1-s) \begin{bmatrix} x(0,0) \\ y(0,0) \\ f(0,0) \end{bmatrix} - (1-r)s \begin{bmatrix} x(0,1) \\ y(0,1) \\ f(0,1) \end{bmatrix} \tag{15}$$

$$- r(1-s) \begin{bmatrix} x(1,0) \\ y(1,0) \\ f(1,0) \end{bmatrix} - rs \begin{bmatrix} x(1,1) \\ y(1,1) \\ f(1,1) \end{bmatrix},$$

where for example,

$$\begin{bmatrix} x(0,s) \\ y(0,s) \\ f(0,s) \end{bmatrix} = 2(\tfrac{1}{2}-s)(1-s) \begin{bmatrix} x(0,0) \\ y(0,0) \\ f(0,0) \end{bmatrix}$$

$$+ 4s(1-s) \begin{bmatrix} x(0,\frac{1}{2}) \\ y(0,\frac{1}{2}) \\ f(0,\frac{1}{2}) \end{bmatrix} + 2s(s-\tfrac{1}{2}) \begin{bmatrix} x(0,1) \\ y(0,1) \\ f(0,1) \end{bmatrix}. \tag{16}$$

derived by making different choices. So for example, the element in Figure 5 is obtained from parametrizing the curves $\begin{bmatrix} x(0,s) \\ y(0,s) \end{bmatrix}$, etc. by quadratics and linear blending these curves together with linear boundary functions f(0,s), etc.. Elements with nodes interior to $\mathcal{E}$ arise when the blending functions are chosen to be polynomials of degree greater than 1. Cf (22).

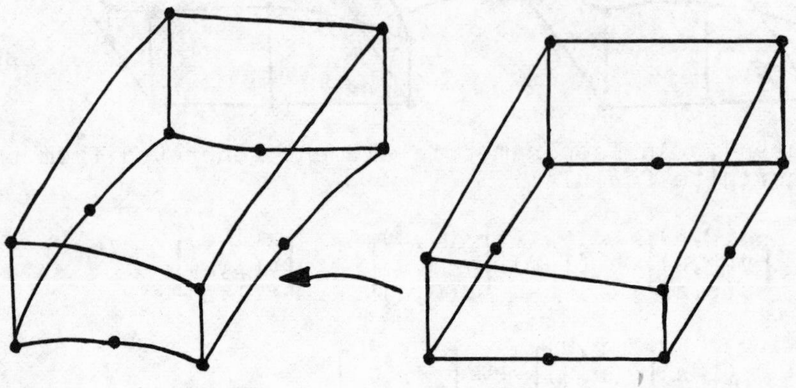

Figure 5. Super-parametric element generated from the linear blending (transfinite) scheme.

Interior Constraints.

We can generalize the above notions in two ways: first, we may consider mappings of $R^2 \to E^n$ for general n; and secondly, we may interpolate $\underline{F}$ not only on the boundary of the region $\mathcal{E} = \{(x_1,x_2,\ldots, x_n)^T = \underline{F}(r,s): 0 \le r,s \le 1\}$, but also along other 'flow lines' or constant generalized co-ordinate lines. To this end, let $0 = r_0 < r_1 < \ldots < r_M = 1$ and $0 = s_0 < s_1 < \ldots < s_N = 1$. Let $\{\phi_i(r)\}_{i=0}^M$ and $\{\psi_i(s)\}_{j=0}^N$ be functions satisfying

$$\phi_i(r_k) = \delta_{ik}, \quad \psi_j(s_\ell) = \delta_{j\ell} \tag{17}$$

$0 \le i, k \le M$, $0 \le j, \ell \le N$, and define the projections (analagous to scalar interpolation)

$$\mathcal{P}_r[\underline{F}] \equiv \sum_{i=0}^{M} \phi_i(r)\underline{F}(r_i,s) \; , \qquad \mathcal{P}_s[\underline{F}] \equiv \sum_{j=0}^{N} \psi_j(s)\underline{F}(r,s_j) \; . \tag{18}$$

The product projection

$$\mathcal{P}_r\mathcal{P}_s[\underline{F}] \equiv \sum_{i=0}^{M}\sum_{j=0}^{N} \phi_i(r)\psi_j(s)\underline{F}(r_i,s_j) \tag{19}$$

interpolates to $\underline{F}$ on the finite point set $\{(r_i,s_j)\}_{i=0,j=0}^{M,N}$ while the Boolean sum or transfinite interpolant

$$(\mathcal{P}_r\oplus\mathcal{P}_s)[\underline{F}] \equiv \mathcal{P}_r[\underline{F}] + \mathcal{P}_s[\underline{F}] - \mathcal{P}_r\mathcal{P}_s[\underline{F}] \tag{20}$$

interpolates to $\underline{F}$ along the $M + N + 2$ lines $r = r_i$, $0 \le i \le M$ and $s = s_j$, $0 \le j \le N$. That is, if $\underline{U}(r,s) \equiv (\mathcal{P}_r\oplus\mathcal{P}_s)[\underline{F}]$, then

$$\underline{U}(r,s_j) = \underline{F}(r,s_j), \quad 0 \le j \le N \; , \qquad \underline{U}(r_i,s) = \underline{F}(r_i,s), \quad 0 \le i \le M \; . \tag{21}$$

If $M = N = 1$ and $r_0 = s_0 = 0$, $r_1 = s_1 = 1$, (20) reduces to the transfinite bilinear interpolant in (12). If $M = N = 2$ and $r_0 = s_0 = 0$, $r_1 = s_1 = \frac{1}{2}$, $r_2 = s_2 = 1$, then

$$\begin{aligned} \phi_0(r) &= 2(r-\tfrac{1}{2})(r-1), & \psi_0(s) &= 2(s-\tfrac{1}{2})(s-1), \\ \phi_1(r) &= 4r(1-r), & \psi_1(s) &= 4s(1-s), \\ \phi_2(r) &= 2r(r-\tfrac{1}{2}), & \psi_2(s) &= 2s(s-\tfrac{1}{2}), \end{aligned} \tag{22}$$

in (20) yields the biquadratically blended interpolant of $\underline{F}$ which matches $\underline{F}$ along the six lines $r = 0,\frac{1}{2},1$ and $s = 0,\frac{1}{2},1$. One can also choose the $\phi_i(r)$ independent of the $\psi_j(s)$. (See Figure 6.)

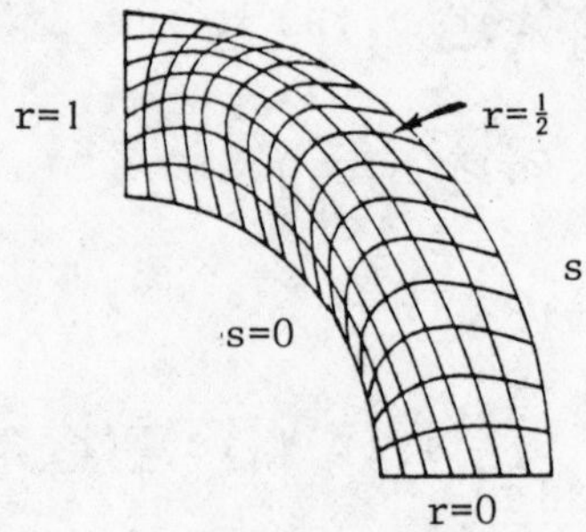

Figure 6. Curvilinear co-ordinate system induced by quadratically blending in r and linearly blending in s.

Another example of the transfinite interpolation formula given in (20) follows from letting the $\phi_i(r)$ and $\psi_j(s)$ be the cardinal functions for natural cubic spline interpolation (see [1, p. 52]) over the knots r_i and s_j, respectively. Then among all functions $\underline{U}$ which satisfy (21), the vector-valued cubic spline-blended interpolant is the 'smoothest' in the sense that it minimizes the functional

$$\int_0^1 \int_0^1 (\underline{U}^{(2,2)})^2 ds\ dt \tag{23}$$

where the integration is interpreted componentwise [7]. Our experience indicates that cubic spline blending functions are very advantageous when several interior generalized co-ordinate lines are specified a priori.

If $\underline{F}$ is a mapping from R^3 into E^3, then the transfinite interpolation formula analogous to (20) is

$$\begin{aligned}(\mathcal{P}_r \oplus \mathcal{P}_s \oplus \mathcal{P}_t)[\underline{F}] \equiv \mathcal{P}_r[\underline{F}] + \mathcal{P}_s[\underline{F}] + \mathcal{P}_t[\underline{F}] - \mathcal{P}_r\mathcal{P}_s[\underline{F}] \\ - \mathcal{P}_r\mathcal{P}_t[\underline{F}] - \mathcal{P}_s\mathcal{P}_t[\underline{F}] + \mathcal{P}_r\mathcal{P}_s\mathcal{P}_t[\underline{F}],\end{aligned} \tag{24}$$

where

$$\begin{aligned}\mathcal{P}_r[\underline{F}] &= \sum_{i=0}^{M} \phi_i(r)\underline{F}(r_i,s,t) ,\\ \mathcal{P}_s[\underline{F}] &= \sum_{j=0}^{N} \psi_j(s)\underline{F}(r,s_j,t) ,\\ \mathcal{P}_t[\underline{F}] &= \sum_{k=0}^{P} \eta_k(t)\underline{F}(r,s,t_k) .\end{aligned} \tag{25}$$

For example, if $\phi_i(r)$ and $\psi_j(s)$ are as in (22) and $\eta_0(t) = 2(t-\frac{1}{2})(t-1)$, $\eta_1(t) = 4t(1-t)$ and $\eta_2(t) = 2t(t-\frac{1}{2})$ then (24) is the <u>triquadratic transfinite interpolant</u> of $\underline{F}$ which matches $\underline{F}$ on the nine planes $r = 0,\frac{1}{2},1$, $s = 0,\frac{1}{2},1$ and $t = 0,\frac{1}{2},1$.

<u>Mesh Generation</u>.

The mapping (20) in two dimensions, or (24) in three dimensions, affords a unique approach to the mesh generation problem for finite element or finite difference applications. Given that $\underline{U}: \mathcal{S} \to \mathcal{E}$ is univalent, we are assured that lines of different generalized co-ordinates $r = \alpha_1$ and $r = \alpha_2$ (or $s = \alpha_1$ and $s = \alpha_2$) will not intersect. Hence if π is a Cartesian product partition of $\mathcal{S}$ determined by the two partitions of the unit interval

$$0 = r_0 < r_1 <\ldots< r_M = 1,\ 0 = s_0 < s_1 <\ldots< s_N = 1,$$

then the generalized co-ordinate lines

$$\{\underline{U}(r_i,s)\}_{i=0}^{M}, \quad\text{and}\quad \{\underline{U}(r,s_j)\}_{j=0}^{N}$$

partition $\mathcal{R}$ into curvilinear quadrilateral elements. By varying the choice of the sets $\{r_i\}_{i=0}^{M}$ and $\{s_j\}_{j=0}^{N}$, one can vary the curvilinear partitioning of $\mathcal{R}$.

Unlike the mesh generator described in [15], we normally do not need to subdivide $\mathcal{R}$ into zones with the boundary segments of each zone parametrized as polynomials. In Figures 7(a)-(c), we illustrate three mesh configurations for the same region $\mathcal{R}$. Each of these utilized the bilinearly blended transfinite map (12); however, the three mesh configurations are basically different. The relative merits of the three partitionings of $\mathcal{R}$ illustrated in Figures 7(a)-(c) must be decided by the engineering analyst in any specific instance.

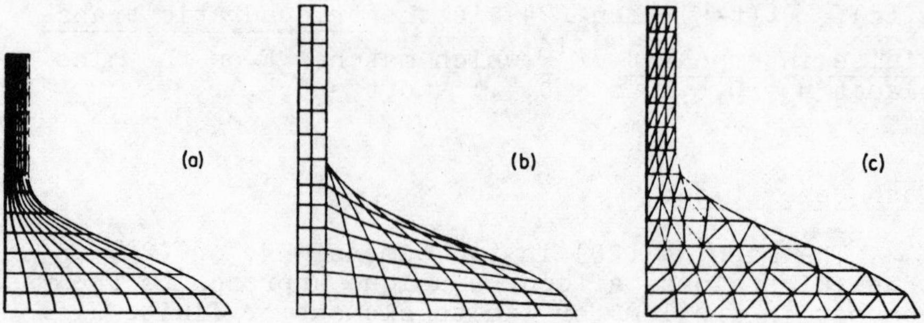

Figure 7. (a) A Cartesian product partitioning of $[0,1] \times [0,1]$ together with the bilinearly blended transfinite map (12) yields a partitioning of $\mathcal{R}$ into curvilinear quadrilateral elements. The configuration is 'rectangular' in that mesh lines do not terminate interior to $\mathcal{R}$. (b) Decomposition of $\mathcal{R}$ into two zones, mapping each zone by (12) and using a Cartesian product partitioning of each zone avoids overly fine partitioning in some areas. The configuration is such that mesh lines do not terminate interior to $\mathcal{R}$. The anomalies in the partitioning of the right-hand zone could be eliminated by reparametrization of the curved boundary of that zone. (c) Recall that the mesh configuration in (a) is the image of a Cartesian product partition of $\mathcal{S}$. Points of that partition were manually deleted to obtain a triangulation of $\mathcal{R}$ in which long slender triangles were avoided.

This process can be envisioned as triangulating $\mathcal{S}$ and then inducing a mesh on $\mathcal{R}$ via the mapping $\underline{U}$.

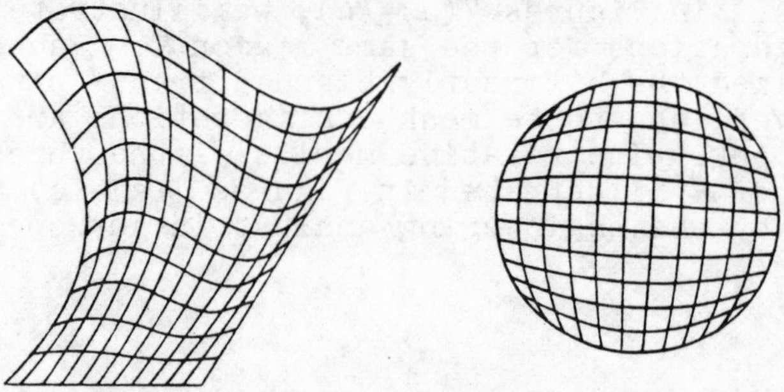

Figure 8. These two regions were co-ordinatized using the bilinearly blended transfinite mapping (12).

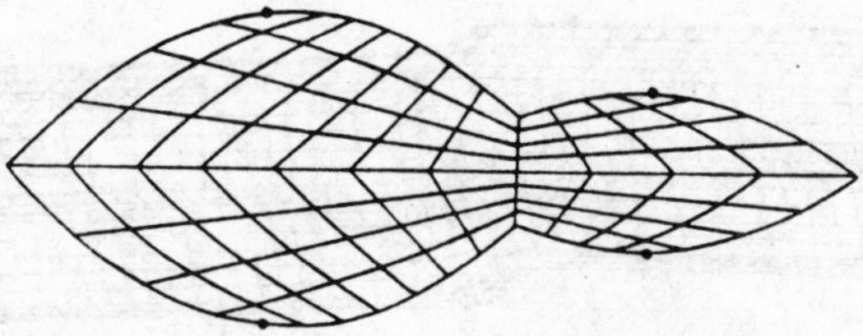

Figure 9. The most natural parametrization of this region (i.e. identification of the four corners of $\mathcal{R}$ as being the images of the corners of $\mathcal{S}$) is known to be such that a univalent map is impossible to construct by means of the technique discussed above. By introducing a straight line constraint at the 'neck' of $\mathcal{R}$ and choosing the four corners as indicated, (free-end or natural) cubic spline blending yields the co-ordinatization shown.

Shells.

The extension of the 'blending-function' mapping or 'transfinite' methods from two-dimensional domains in E^2 to simply connected shells in E^3 is straightforward. Instead of dealing with plane curves, we consider the functions $\underline{F}(r_i,s)$ and $\underline{F}(r,s_j)$ in the generic expression (20) to be vector-valued functions of three (instead of two) components, so that the graph of the vector-valued function

$$\underline{U}(r,s) \equiv \begin{bmatrix} X(r,s) \\ Y(r,s) \\ Z(r,s) \end{bmatrix}$$

is a bounded, simply connected shell in Euclidean 3-space.

As a practical matter, draftsmen represent space curves by providing two sets of plane projected views of the curve in 3-space, for example, one is provided with views in any two of the following orthogonal planes: (x,y), (x,z) or (y,z). By electronically digitizing and computationally 'marrying' any two planar views of a set of key design curves one can obtain the necessary pointwise representation of the three component vector-valued functions needed in expression (20).

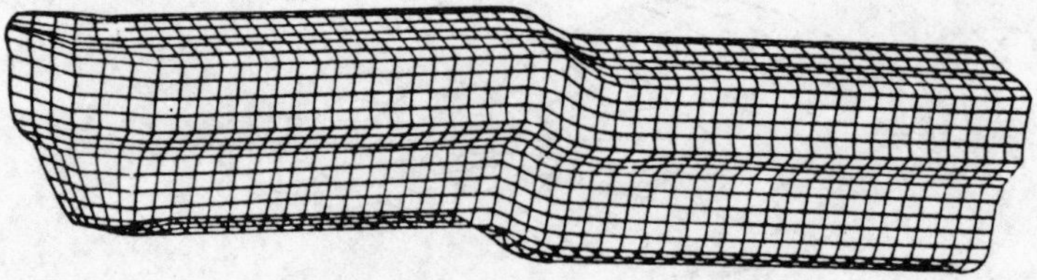

Figure 10. Example of the application of blending function methods to shell decomposition.

As an example, the boundary curves and certain principal feature lines were digitized from a set of engineering drawings of the exterior shell of half of an automobile front bumper. Using cubic splines as the blending functions in (20) one obtains a complete mathematical description of the bumper surface as a map from $\mathcal{S} = [0,1] \times [0,1]$ onto the surface in three space. Figure 10 shows the curvilinear co-ordinatization induced on the surface as the image of a Cartesian product decomposition of the unit square $\mathcal{S}$. The application to mesh generation is now obvious.

Global Inversion.

In [4] we investigate conditions under which a mapping such as (20) is a bijection from $\mathcal{S}$ to the closed, bounded set $\mathcal{E}$ having $\bar{x}(\partial\mathcal{S})$ as its boundary. In particular, we consider the cases when the curves $\bar{x}(0,s)$, $\bar{x}(1,s)$, $\bar{x}(r,0)$ and $\bar{x}(r,1)$ are either four straight line segments specified by the four nodes (points) $\bar{x}(i,j)$, $i,j = 0,1$, or four parabolic arcs specified by the eight nodes $\bar{x}(i,j)$, $\bar{x}(\frac{1}{2},j)$, $\bar{x}(i,\frac{1}{2})$, $i,j = 0,1$. Then (20) reduces respectively, to the well-known bilinear or quadratic isoparametric transformations (14) of finite element analyses, and $\mathcal{E}$ is known as the four- or eight-node isoparametric element [16] (see Figure 11).

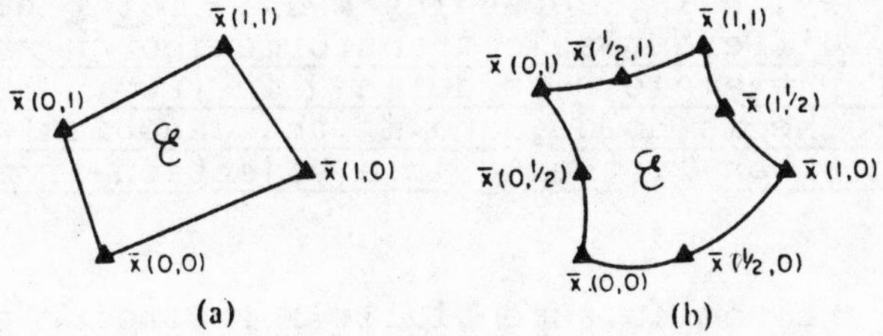

Figure 11. (a) 4-node isoparametric element. (b) 8-node isoparametric element.

Considerations concerning the bijective nature of isoparametric transformations are important from both the theoretical and practical points of view. For instance, the numerical solution of boundary value problems by finite element techniques employing isoparametric elements requires the evaluation of certain integrals by means of the change of variables defined by $\underline{U}$. Thus, knowledge of the bijectivity of $\underline{U}$ is necessary to insure that this change of variables is in fact proper. Furthermore, after the isoparametric finite element solution has been found, the actual inversion of (14) is necessary to obtain values of the dependent variables, such as stress, at prescribed points of $\mathcal{E}$. Therefore, in addition to establishing the a priori existence of an inverse of $\underline{U}$, it is also useful to have an algorithm for its pointwise inversion. Such an algorithm is given in [4,5].

Clearly, if $\bar{x}$: $\partial\mathcal{S} \to R^2$ is not an injection, then $\underline{U}$: $\mathcal{S} \to R^2$ as defined by (14) cannot be a bijection to any set having $\bar{x}(\partial\mathcal{S})$ as its boundary. Therefore, we state the following fundamental

Boundary Hypothesis: The continuous transformation $\bar{x}$: $\partial\mathcal{S} \to R^2$ is an injection.

This condition is obviously equivalent to hypothesizing that $\bar{x}(\partial\mathcal{S})$ is a simple closed curve. Under the boundary hypothesis, we know from the Jordan Curve Theorem that $\bar{x}(\partial\mathcal{S})$ partitions the plane into two disjoint, open, connected sets and forms their common boundary. Furthermore, only one of these sets is bounded and in the sequel it is the closure of this bounded set that we take as the set or "element" $\mathcal{E}$.

THEOREM 1 (Frey, Hall, Porsching [4]). *Let $\underline{U}$ be a continuously differentiable transformation on an open set* $T \supset \mathcal{S}$, *for example* $\underline{U}$ *as defined in (20). If the boundary hypothesis holds, and if the Jacobian of* $\underline{U}$ *does not vanish on* T, *then* $\underline{U}$ *is a bijection from* $\mathcal{S}$ *to* $\mathcal{E}$.

Proof. The theorem is essentially a rewording of a result of de la Vallée Poussin [13, p. 355], and the reader is referred to this reference for the details of the proof. Q.E.D.

Bilinear Transformations.

The bilinear isoparametric transformation results from (14) when the four nodes $\bar{x}(i,j)$, $i,j = 0,1$, are given and $\bar{x}(\partial\mathcal{S})$ is defined by

$$\bar{x}(0,s) \equiv (1-s)\bar{x}(0,0) + s\bar{x}(0,1),$$
$$\bar{x}(1,s) \equiv (1-s)\bar{x}(1,0) + s\bar{x}(1,1),$$
$$\bar{x}(r,0) \equiv (1-r)\bar{x}(0,0) + r\bar{x}(1,0),$$
$$\bar{x}(r,1) \equiv (1-r)\bar{x}(0,1) + r\bar{x}(1,1).$$

In this case, if we denote the left side of (14) by $\underline{T}_1(r,s)$, **then**

THEOREM 2 (Frey, Hall, Porsching [4]). *Consider the transformation* $\underline{T}_1$ *and assume that the boundary hypothesis holds. Then the following conditions are equivalent*:

(i) *The four-node isoparametric element* $\mathcal{E}$ *is convex*

(ii) *The Jacobian of* $\underline{T}_1$ *is positive at the four vertices of* (i.e. $(r,s) = (i,j)$, $i,j = 0,1$).

(iii) $\underline{T}_1$ *is a bijection from* $\mathcal{S}$ *to* $\mathcal{E}$.

Quadratic Transformations.

Now suppose that the eight nodes $\bar{x}(i,j)$, $\bar{x}(\tfrac{1}{2},j)$, $\bar{x}(i,\tfrac{1}{2})$, $i,j = 0,1$, are given; cf. Figure 11(b). We define $\bar{x}(\partial\mathcal{S})$ by

$$\begin{aligned}
\bar{x}(0,s) &\equiv 2(s-\tfrac{1}{2})(s-1)\bar{x}(0,0)-4s(s-1)\bar{x}(0,\tfrac{1}{2})+2s(s-\tfrac{1}{2})\bar{x}(0,1),\\
\bar{x}(1,s) &\equiv 2(s-\tfrac{1}{2})(s-1)\bar{x}(1,0)-4s(s-1)\bar{x}(1,\tfrac{1}{2})+2s(s-\tfrac{1}{2})\bar{x}(1,1),\\
\bar{x}(r,0) &\equiv 2(r-\tfrac{1}{2})(r-1)\bar{x}(0,0)-4r(r-1)\bar{x}(\tfrac{1}{2},0)+2r(r-\tfrac{1}{2})\bar{x}(1,0),\\
\bar{x}(r,1) &\equiv 2(r-\tfrac{1}{2})(r-1)\bar{x}(0,1)-4r(r-1)\bar{x}(\tfrac{1}{2},1)+2r(r-\tfrac{1}{2})\bar{x}(1,1).
\end{aligned} \qquad (26)$$

When this is used in conjunction with (14) the resulting transformation, which we denote by $\underline{T}_2(r,s)$, is called the 8-node quadratic isoparametric transformation. Of course, Theorem 1 again applies. However, we have been unable to find an analogue of Theorem 2 relating bijectivity directly to an obvious geometric property of the set $\mathcal{E}$. We do have the following specific results however.

Let $\bar{P}_i = (x_i, y_i)$, $i = 1,\ldots, 8$, denote the given nodes, where $\bar{x}(0,0) = \bar{P}_1$, $\bar{x}(1,0) = \bar{P}_2$, $\bar{x}(1,1) = \bar{P}_3$, $\bar{x}(0,1) = \bar{P}_4$, $\bar{x}(\tfrac{1}{2},0) = \bar{P}_5$, $\bar{x}(1,\tfrac{1}{2}) = \bar{P}_6$, $\bar{x}(\tfrac{1}{2},1) = \bar{P}_7$ and $\bar{x}(0,\tfrac{1}{2}) = \bar{P}_8$. We consider a special class of quadratic isoparametric transformations obtained by requiring that the boundary transformation $\bar{x}$ satisfy:

(a) $\bar{x}(r,0) = (x_2 r, 0)$,

(b) $\bar{x}(0,s) = (0, y_4 s)$,

(c) under componentwise ordering, $\bar{x}(r,1) \geq (0,\epsilon)$ $\bar{x}(1,s) \geq (\epsilon,0)$ for some $\epsilon > 0$,

(d) $x_7 = x_3/2$ and $y_6 = y_3/2$.

If (a)-(d) and the boundary hypothesis hold, we call the set $\mathcal{E}$ a semirectangle. Figure 12(a) shows a typical semirectangular element. Conditions are given in [4] which guarantee $\underline{T}_2$ is a bijection for semirectangles. In particular, <u>if two sides are parallel lines</u> (Figure 12b) <u>then $\underline{T}_2$ is a bijection</u>.

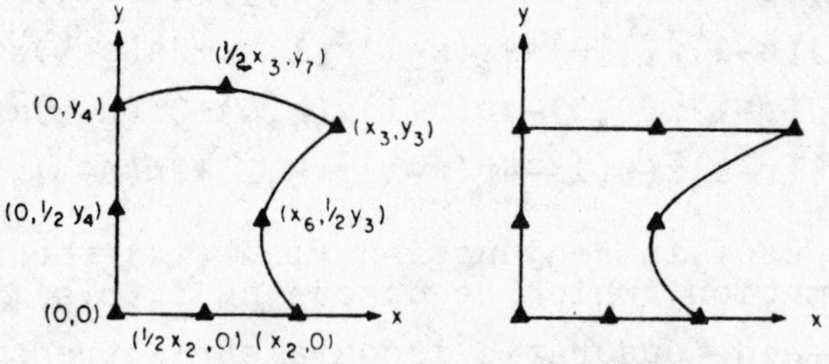

(a) (b)

Figure 12. Semirectangles

Let the convex quadrilateral Q have vertices and side midpoints $\bar{Q}_i$, i = 1,..., 8, shown in Figure 13. Then the associated quadratic transformation defined by (14) and (26) and the nodes $\bar{Q}_i$, i = 1,..., 8, is in fact bilinear and by Theorem 2 has a positive Jacobian on $\mathcal{S}$. A nondegenerate quadratic transformation may be obtained by perturbing the midside nodes from ∂Q. By continuity, the Jacobian remains positive for all sufficiently small perturbations. In [4] the following bounds on the size of perturbations which guarantee that the associated transformation has a positive Jacobian on $\mathcal{S}$ were developed.

Suppose that the transformation T_2 is defined by the nodes $\bar{P}_i = (x_i, y_i)$, where $\bar{P}_i = \bar{Q}_i$, i = 1,..., 4, $P_i = \bar{Q}_i + \bar{\eta}_i$, i = 5,..., 8. We consider the class of perturbations $\bar{\eta}_i$ for which $\bar{\eta}_5 = (0, \eta_5)$, $\bar{\eta}_6 = (\eta_6, 0)$, $\bar{\eta}_7 = (0, \eta_7)$, $\bar{\eta}_8 = (\eta_8, 0)$, and we assume that $x_1 < x_2$, $x_4 < x_3$, $y_1 < y_4$, $y_2 < y_3$. See Figure 13.

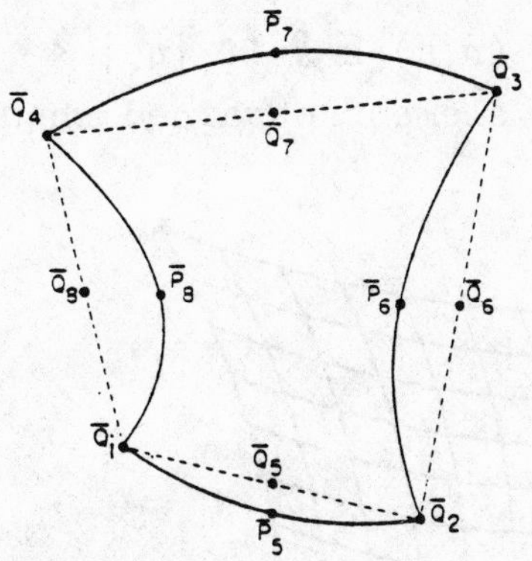

Figure 13. Quadratic element obtained by perturbations

THEOREM 3 (Frey, Hall, Porsching [4]). *If an 8-node element* $\mathcal{E}$ *is obtained from a convex quadrilateral Q by perturbations of its midside nodes in the manner shown in Figure 13, and if*

$$|n_i| \leq \frac{-B_1 + \sqrt{B_1^2 + 4m}}{8}, \quad i = 5, \dots, 8, \tag{27}$$

where B_1 *and* m *are given by*

$$B_1 = [|x_{21}|+|y_{32}|+|x_{34}|+|y_{41}|]/4 + 2\max\{|x_{41}|,|x_{32}|\} + 2\max\{|y_{21}|,|y_{34}|\}, \tag{28a}$$

$$m = \min[y_{41}x_{21}-x_{41}y_{21}, y_{32}x_{21}-x_{32}y_{21}, y_{41}x_{34}-x_{41}y_{34}, y_{32}x_{34}-x_{32}y_{34}], \tag{28b}$$

and

$$x_{ij} = x_i - x_j \quad \text{and} \quad y_{ij} = y_i - y_j,$$

then the Jacobian of the associated quadratic transformation $\underline{T}_2$ *is positive on* $\mathcal{S}$. *Furthermore, if the boundary hypothesis holds, then* $\underline{T}_2$ *is a bijection from* $\mathcal{S}$ *to* $\mathcal{E}$.

Suppose, for example, that Q is defined by the nodes $\bar{Q}_1 = (0,0)$, $\bar{Q}_2 = (1,0)$, $\bar{Q}_3 = (1.3,1.2)$, $\bar{Q}_4 = (-.2,1)$. Then $B_1 = 2.175$, $m = 1$; and therefore, det. $J(\epsilon) > 0$ for all $(r,s) \in \mathcal{S}$ if $|\eta_i| < .09747$. Figure 14a shows the element obtained when $\eta_i = -.097$, $i = 5,6,7,8$.

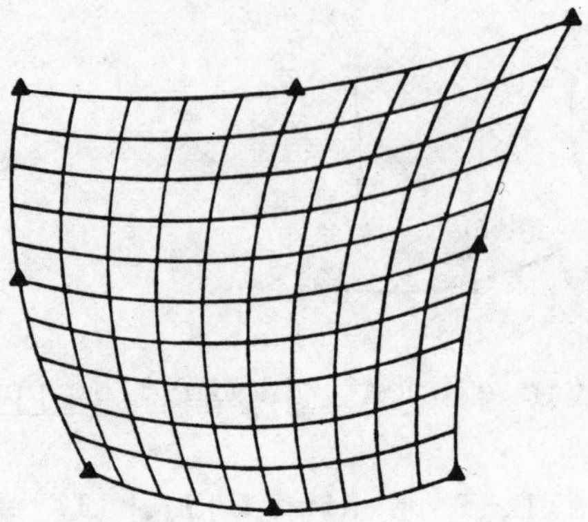

8 NODE 2-D ELEMENT

Figure 14a. Perturbed element with the associated invertible transformation $\underline{T}_2$. The images of $s = i/10$ and $r = i/10$, $1 \leq i \leq 10$ are also displayed.

It can be shown that choosing $\eta_i = -0.27$, $i = 5,6,7,8$ yields a noninvertible map. A more dramatic example of noninvertibility is shown in Figure 14b. This map results from the choice $\eta_5 = \eta_6 = -\eta_7 = .5$ and $\eta_8 = .1$.

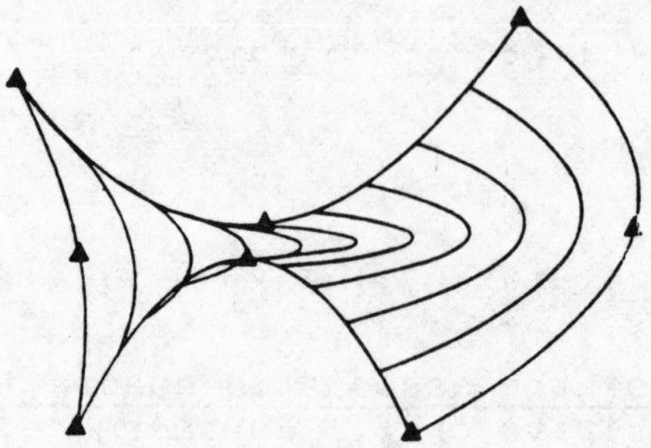

8 NODE 2-D ELEMENT

Figure 14b. Perturbations too large in magnitude. The associated transformation $\underline{T}_2$ is noninvertible.

Overspill. The term overspill has been used to describe instances when $\underline{T}: \mathcal{S} \to R^2$ is such that $\underline{T}(\mathcal{S})$ properly contains $\mathcal{E}$, a specified set (element), [16]. When such is the case, the image of a constant coordinate line, for example s = s* originates at one boundary curve of $\mathcal{E}$, say at $\bar{x}(0,s^*)$ extends "beyond" $\bar{x}(1,s)$ and returns by design to terminate at $\bar{x}(1,s^*)$; it overspills the set $\mathcal{E}$. In other words, the image of some constant coordinate line intersects $\partial\mathcal{E}$ in more than two points. Formally, we say that the transformation $\underline{T}$ of (14) has the no overspill property if $\underline{T}\bar{z} \notin \bar{x}(\partial\mathcal{S})$ when $\bar{z} \in \mathcal{S}^o$, where $\mathcal{S}^o$ denotes the interior of $\mathcal{S}$.

LEMMA 1 [4]. Let the boundary hypothesis hold, and let $\underline{T}$ have the no overspill property. Then $\underline{T}(\mathcal{S}) \subseteq \mathcal{E}$.

It is clear that no overspill is a necessary condition for $\underline{T}$ to be a bijection of $\mathcal{S}$ to an element $\mathcal{E}$ having $\bar{x}(\partial\mathcal{S})$ as its boundary. It is also sufficient for a large subclass of the quadratic isoparametric transformation $\underline{T}_2$ defined by (14) and (26). We now define this subclass.

Denote the eight nodes appearing in (26) by $\bar{P}_i = (x_i,y_i)$, i = 1,..., 8. Now suppose that they satisfy (cf. Figure 13)

Assumption A1.

$$x_5 = \tfrac{1}{2}(x_1+x_2),\ x_7 = \tfrac{1}{2}(x_3+x_4),\ x_1 \neq x_2,\ x_4 \neq x_3,$$

$$y_6 = \tfrac{1}{2}(y_2+y_3),\ y_8 = \tfrac{1}{2}(y_1+y_4),\ y_1 \neq y_4,\ y_2 \neq y_3,$$

and let the boundary hypothesis hold. The boundary of $\mathcal{E}$ then consists of four parabolic arcs, two of which have the generic functional form $y = f(x) \equiv ax^2 + bx + c$, and two of which have the form $x = g(y) \equiv Ay^2 + By + C$. See Figure 15. As the following lemma establishes, this is also true of the images of the r and s coordinate lines.

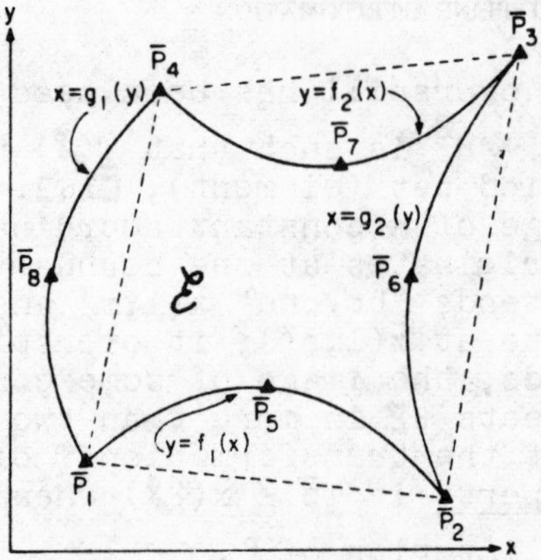

Figure 15. An 8-node element with parabolic boundary segments.

LEMMA 2 [4]. *Under Assumption A1, the curve* $\underline{T}_2(r,s^*) \equiv (x(r,s^*),y(r,s^*))$ *is a parabola* $y = f(x)$ *with axis parallel to the y-axis for each fixed* $s = s^*$, *and* $\underline{T}_2(r^*,s)$ *is a parabola* $x = g(y)$ *with axis parallel to the x-axis for each fixed* $r = r^*$.

The final assumption that we need to define the subclass concerns the rates of change of the tangents of a typical pair of parabolas $y = f(x)$ and $x = g(y)$ appearing in Lemma 2. Specifically, we assume that

$$\max_{(x,f(x))\in\mathcal{E}} |f''(x)| < \min_{(x,g^{-1}(x))\in\mathcal{E}} |[g^{-1}(x)]''|.$$

In terms of the transformation $\underline{T}_2(r,s) \equiv (x(r,s),y(r,s))$, this becomes

Assumption A2.

$$\max_{0\le r,s\le 1}\left|\frac{\partial^2 y}{\partial r^2}\left(\frac{\partial x}{\partial r}\right)^{-3}\right| < \min_{0\le r,s\le 1}\left|\frac{\partial^2 x}{\partial s^2}\left(\frac{\partial x}{\partial s}\right)^{-3}\right|.$$

Note that $\partial^2 y/\partial r^2$ and $\partial^2 x/\partial s^2$ are respectively independent of r and s. Moreover, if Assumption A1 holds, then $\partial x/\partial r$ also does not depend on r, and the left side of the above inequality is a function of s only.

THEOREM 4 (Frey, Hall, Porsching [4]). *Let the transformation* $\underline{T}_2$ *be defined by* (14) *and* (26). *Assume that the boundary hypothesis holds and that Assumptions A1 and A2 are true. If* $\underline{T}_2$ *has the no overspill property, then it is a bijection from* $\mathcal{S}$ *to* $\mathcal{E}$.

Some remarks on Assumption A2 are now in order. In the first place, if A1 holds, then we find by direct computation, using (14) and (26), that

$$
\begin{aligned}
\partial x/\partial r &= x(1,s) - x(0,s), \\
\partial^2 y/\partial r^2 &= 4[(1-s)(y_1-2y_5+y_2)+s(y_4-2y_7+y_3)], \\
\partial x/\partial s &= (1-r)[s(4x_1-8x_8+4x_4)-(3x_1-4x_8+x_4)] \\
&\quad + r[s(4x_2-8x_6+4x_3)-(3x_2-4x_6-x_3)], \\
\partial^2 x/\partial s^2 &= 4[(1-r)(x_1-2x_8+x_4)+r(x_2-2x_6+x_3)].
\end{aligned}
$$

Therefore, for A2 to hold, it is necessary that $x_1 - 2x_8 + x_4$ and $x_2 - 2x_6 + x_3$ have the same sign. That is, the parabolas $\underline{T}_2(0,s)$ and $\underline{T}_2(1,s)$ should both open to the right or left (See Figure 15). When this is the case, it is easy to see that the right side of the inequality in A2 is bounded below by

$$m \equiv \frac{4\min(|x_1-2x_8+x_4|,\ |x_2-2x_6+x_3|)}{d^3} \tag{29}$$

where

$$
\begin{aligned}
d = \max(&|4x_6-3x_2-x_3|, |4x_6-3x_3-x_2|, |4x_8-3x_1-x_4|, \\
&|4x_8-3x_4-x_1|),
\end{aligned}
$$

and the left side is bounded above by

$$M \equiv \frac{4\max(|y_1-2y_5+y_2|,\ |y_4-2y_7+y_3|)}{[\min\limits_{0\le s\le 1} |x(1,s)-x(0,s)|]^3}. \tag{30}$$

Note that in any given case, it is a simple matter to obtain the quantities m and M, the denominator in M giving rise to an elementary minimization problem via (26). Clearly, A2 holds if $M < m$, which is certainly the case if $\underline{T}_2(r,0)$ and $\underline{T}_2(r,1)$ are straight line segments since then $M = 0$.

In view of Theorem 4, we now seek conditions which guarantee that quadratic transformations $\underline{T}_2$

satisfying A1 will have the no overspill property. Consider the element $\mathcal{E}$ in Figure 13. If both coordinates of the four midside nodes were averaged, $\mathcal{E}$ would be the straight sided quadrilateral Q with vertices $\bar{Q}_i$, i = 1,2,3,4 as indicated by the dotted lines in Figure 13. In this case $\underline{T}_2 \equiv \underline{T}_1$ and Theorem 2 applies. The element $\mathcal{E}$ differs from Q by perturbations η_6 and η_8 in the x-coordinate of $\bar{Q}_6$ and $\bar{Q}_8$, and perturbations η_5 and η_7 in the y-coordinate of $\bar{Q}_5$ and $\bar{Q}_7$. The question is: <u>How large can these perturbations be without producing overspill?</u> Before answering this question we find it necessary to make a further assumption.

<u>Assumption A3</u>. Assume that for each r = r* (resp. s = s*) the straight line segment

$$\begin{aligned}&(1-s)\bar{x}(r^*,0) + s\bar{x}(r^*,1), \quad 0 \leq s \leq 1\\&(\text{resp. } (1-r)\bar{x}(0,s^*) + r\bar{x}(1,s^*), \quad 0 \leq r \leq 1)\end{aligned} \tag{31}$$

intersects each of the boundary curves $\bar{x}(r,0)$ and $\bar{x}(r,1)$ (resp. $\bar{x}(i,s)$, i = 0,1) once and only once.

If η_6 and η_8 (resp. η_5 and η_7) are zero, then A3 guarantees that $\underline{T}_2$ is one-to-one since $\underline{T}_2$ is then just a "railing" of the curves $\bar{x}(r,0)$ and $\bar{x}(r,1)$, (resp. $\bar{x}(0,s)$ and $\bar{x}(1,s)$). That is,

$$\begin{aligned}&\underline{T}_2(r,s) = \underline{P}_s(r,s) \equiv (1-s)\bar{x}(r,0) + s\bar{x}(r,1)\\&(\text{resp. } \underline{T}_2(r,s) = \underline{P}_r(r,s) \equiv (1-r)\bar{x}(0,s) + r\bar{x}(1,s)).\end{aligned} \tag{32}$$

Conversely, if $\underline{P}_r$ and $\underline{P}_s$ are invertible for $0 \leq r, s \leq 1$, then Assumption A3 holds, hence we have

LEMMA 3. <u>Assumption</u> A3 <u>holds if and only if</u> $\underline{P}_r$ <u>and</u> $\underline{P}_s$ <u>are injections on</u> $\mathcal{S}$.

The importance of Lemma 3 is that the validity of A3 can be established computationally by investigating the invertibility of $\underline{P}_r$ and $\underline{P}_s$. By Theorem 1, we need only establish the nonvanishing of their Jacobians. But, the Jacobian of $\underline{P}_s$ is by direct calculation of the form $(1-s)q_1(r) + sq_2(r)$, $0 \leq s \leq 1$, where $q_i(r)$, $i = 1,2$, are quadratics in r. The non-vanishing of the Jacobian is then established by checking if $q_1(r)q_2(r) > 0$ for $0 \leq r \leq 1$.

In [4] bounds on the perturbations η_5, η_6, η_7 and η_8 are determined so as to guarantee that $\underline{T}_2$ has the no overspill property.

THEOREM 5 (Frey, Hall, Porsching [4]). <u>Assume that the boundary hypothesis</u>, A1 <u>and</u> A3 <u>hold and</u>, <u>referring to Figure 15</u>, <u>let</u>

$$x_L = \min_{0\leq s\leq 1} x(0,s),\quad x_R = \max_{0\leq s\leq 1} x(1,s),$$

$$y_T = \max_{0\leq r\leq 1} y(r,1),\quad y_B = \min_{0\leq r\leq 1} y(r,0),$$

$$M_x = \max\left\{\max_{x_L\leq x\leq x_R} |df_1/dx|,\ \max_{x_L\leq x\leq x_R} |df_2/dx|\right\},$$

$$M_y = \max\left\{\max_{y_B\leq y\leq y_T} |dg_1/dy|,\ \max_{y_B\leq y\leq y_T} |dg_2/dy|\right\},$$

$$S_x = \max_{0\leq r\leq 1} |x(r,1)-x(r,0)|,\quad S_y = \max_{0\leq s\leq 1} |y(1,s)-y(0,s)|,$$

$$H_x = \min_{0\leq s\leq 1} (x(1,s)-x(0,s)),\quad H_y = \min_{0\leq r\leq 1} (y(r,1)-y(r,0)),$$

$$y_5 = (y_1+y_2)/2 + \eta_5,\quad y_7 = (y_3+y_4)/2 + \eta_7,$$

$$x_6 = (x_2+x_3)/2 + \eta_6, \text{ and } x_8 = (x_1+x_4)/2 + \eta_8.$$

<u>If</u>

$$\begin{aligned}\eta_x &\equiv \max\{|\eta_6|,|\eta_8|\} \leq \tfrac{1}{4}\{H_y/M_x - S_x\},\\ \eta_y &\equiv \max\{|\eta_5|,|\eta_7|\} \leq \tfrac{1}{4}\{H_x/M_y - S_y\},\end{aligned} \tag{33}$$

<u>then</u> $\underline{T}_2$ <u>has no overspill property</u>.

Let's consider an example: Let $\bar{P}_1$: (0,0), $\bar{P}_2$: (6,1), $\bar{P}_3$: (5,5), and $\bar{P}_4$: (0,4) be the corner nodes and

$\bar{P}_6$: (5,3) and $\bar{P}_8$: (-0.5,2) be two of the midside nodes of a given element. We consider how the straight lines $\overline{P_1P_2}$ and $\overline{P_3P_4}$ can be deformed into parabolas so as to guarantee that the element does not have the overspill property. A class of such parabolas is described in terms of the perturbations η_5 and η_7 of the y-coordinates of the midside nodes $\bar{P}_5 = (3,0.5+\eta_5)$ and $\bar{P}_7 = (2.5,4.5+\eta_7)$. Note $\eta_5 = \eta_7 = 0$ corresponds to $\underline{T}_2(r,s) = \underline{P}_r(r,s)$, which is one-to-one. We use Theorem 5 to bound $|\eta_5|$ and $|\eta_7|$ as follows:

1. Compute

$$S_x = \max\{|x_4-x_1|, |x_3-x_2|\} = 1,$$
$$S_y = \max\{|y_1-y_2|, |y_4-y_3|\} = 1.$$

2. By direct calculation, using (26),

$$H_x = \min_{0\le s\le 1} |x(1,s)-x(0,s)| = 5.$$

3. We choose $\eta_7 < 0$ and $\eta_5 > 0$, so $y_B = 0$ and $y_T = 5$. By the chain rule

$$\max_{0\le y\le 5} |dg_1/dy| = .75, \quad \max_{0\le y\le 5} |dg_2/dy| = 1.0,$$

so $M_y = 1$.

From (33) $\underline{T}_2$ has the no overspill property if η_5 and η_7 are chosen in magnitude less than 1.0. Figure 16 illustrates this extreme case where $\bar{P}_7 = (2.5,3.5)$ and $\bar{P}_5 = (3.0,1.5)$.

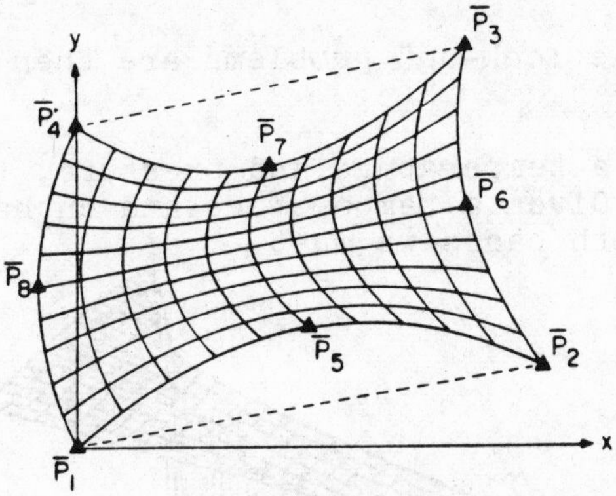

Figure 16. A one-to-one isoparametric mapping $\underline{T}_2$.

Referring back to (29) and (30), we see that $m = 4/27$ and $M = 8/125 < m$. Hence, A2 holds and by Theorem 3, for the element in Figure 16, $\underline{T}_2$ is a bijection from $\mathcal{S}$ to $\mathcal{E}$.

Storage of Tabular Data. As a final application of vector-valued interpolation consider the following: In a real time environment or simulation it is often desirable to have the capability of retrieving good approximate values of tabulated data, e.g. thermodynamic properties of steam and water are necessary in the operation of nuclear and fossil fired power plants. In [11], it was determined that such thermodynamic properties (enthalpy, specific volume, etc.) can best be considered as surfaces constructed over a domain in the temperature-pressure plane (cf. Figure 17) and their approximation reduces to a surface fitting problem. However, there are three important distinctions: (1) the "surface" or tabular data is fixed, (2) the domain is curved and (3) the size of computer (25-30K) involved in process control limits the amount of information that can be stored and efficiently retrieved.

Transfinite interpolation was used in [11] to construct domain transformations and to interpolate steam and water properties. Note that if enthalpy is considered as the dependent variable and pressure-temperature as the independent variables then the domain is as illustrated in Figure 18; similarly, when pressure is chosen as the dependent variable, the domain is as illustrated in Figure 19. Now for either of these domains the transfinite (vector-valued) interpolation scheme $\underline{\breve{U}}$ in (13) induces a curvilinear coordinate system as indicated in Figures 18 and 19. The associated "table look-up" problems are then respectively:

(1) Given a temperature and pressure, find enthalpy, <u>or</u> (2) Given a temperature and enthalpy, find pressure. In both cases we must

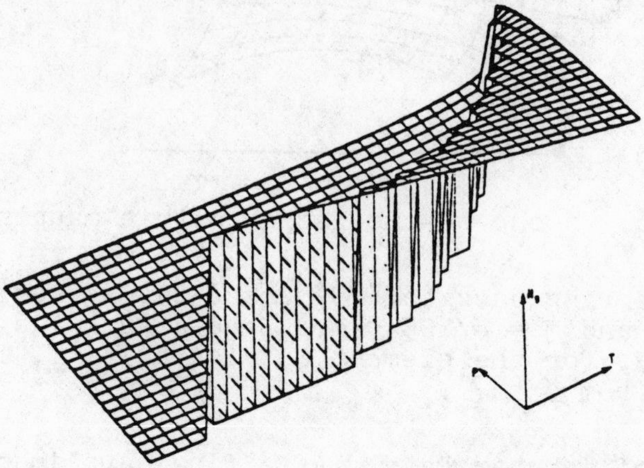

Figure 17. Enthalpy of superheated steam versus temperature and pressure.

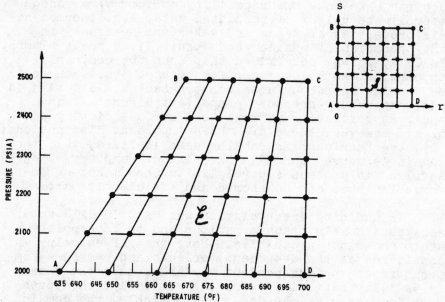

Figure 18. Projection into temperature-pressure plane: enthalpy the dependent variable. The saturation line AB is curved.

in fact invert the map $\underline{U}$ and then evaluate, for example, a linear blended surface (13) at the appropriate (r,s) value.

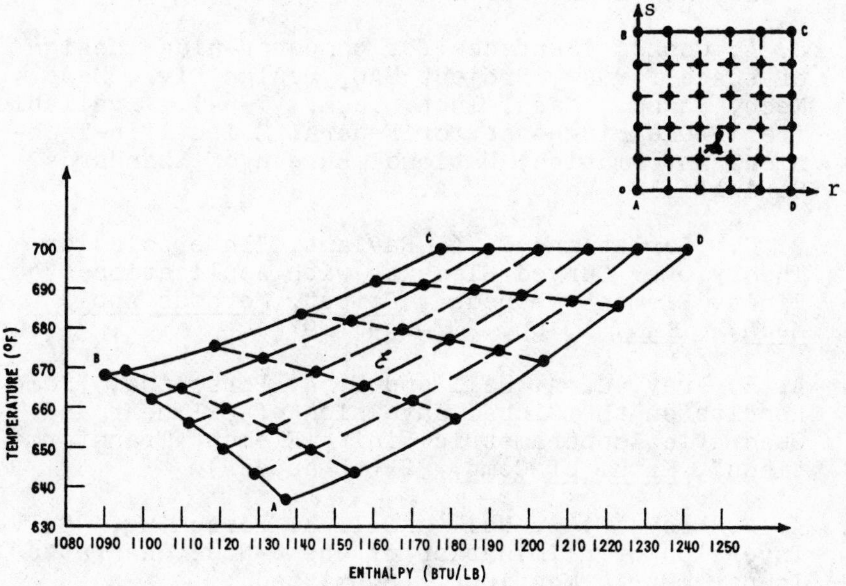

Figure 19. Projection into enthalpy-temperature plane: pressure the dependent variable.

REFERENCES

1. J. H. Ahlberg, E. N. Nilson and J. L. Walsh, The Theory of Splines and Their Applications, Academic Press, New York, 1967.

2. S. A. Coons, 'Surfaces for computer-aided design of space forms', Project MAC, Design Div., Dept. Mech. Engng., Mass. Inst. Tech. (1964). Available from: Clearinghouse for Federal Scientific-Technical Information, National Bureau of Standards, Springfield, Va., U.S.A.

3. P. G. Ciarlet and P. A. Raviart, "Interpolation Theory Over Curved Elements with Applications to Finite Element Methods", Comput. Methods Appl. Mech. Enng. 1, pp. 217-249 (1972).

4. A. E. Frey, C. A. Hall and T. A. Porsching, "Some Results on the Global Inversion of Bilinear and Quadratic Isoparametric Finite Element Transformations", Math. of Comp. 32 (to appear).

5. A. E. Frey, C. A. Hall and T. A. Porsching, "SOLV8: Inversion by Elimination of the 8-Node Quadratic Isoparametric Mapping", (submitted).

6. W. J. Gordon, 'Free-form surface interpolation through curve networks', Res. Rept. 921, General Motors, Warren, Mich., U.S.A. (1969).

7. W. J. Gordon, 'Spline-blended surface interpolation through curve networks', J. Math. Mech., 18, 931-952 (1969).

8. W. J. Gordon, 'Blending-function methods for bivariate and multivariate interpolation and approximation', SIAM J. Numer. Anal. 8, 158-177 (1971).

9. W. J. Gordon and C. A. Hall, "Construction of Curvilinear Co-ordinate Systems and Applications to Mesh Generation", Int. J. for Numer. Methods in Enng. 7, 461-477 (1973).

10. W. J. Gordon and C. A. Hall, 'Transfinite element methods: Blending-function interpolation over curved element domains', Numerishe Mathematik 21, 109-129 (1973).

11. C. A. Hall and B. A. Mutafelija, "Transfinite Interpolation of Steam Tables", J. Computational Physics 18, 79-91 (1975).

12. P. Morse and H. Feshbach, Methods of Theoretical Physics, Pt. I, McGraw-Hill, New York, 1953.

13. C. J. de la Vallee Poussin, Cours d'Analyse Infinitesimale, Vol. I, Gauthier-Villars, Paris, 1926.

14. D. S. Watkins and P. Lancaster, "Some Families of Finite Elements", J. Inst. Math. Appl. 19, 385-397 (1977).

15. O. C. Zienkiewicz and D. V. Phillips, 'An automatic mesh generation scheme for plane and curved surfaces by "isoparametric" coordinates', Int. J. Num. Meth. Engg. 3, 519-528 (1971).

16. O. C. Zienkiewicz, The Finite Element Method in Engineering Science, McGraw-Hill, New York, 1971.

THE CONSTRUCTION OF A MACRO ELEMENT FOR USE IN THREE DIMENSIONAL FRACTURE MECHANICS

Charles Hall*

Institute for Computational Mathematics
and Applications
Department of Mathematics and Statistics
University of Pittsburgh

ABSTRACT. The stiffness derivative technique is used to compute stress intensity factors. The crack region is modelled by means of a macro element which contains, a priori, a high density of nodes in the vicinity of the crack front and is compatible with standard 20 node elements that it abutts.

INT. As a specific example of an application of vector valued blended interpolation we consider the following approximation problem which is discussed in more detail in [6,7].

Quantification of the margin of safety against brittle fracture is one of the primary objectives in establishing the structural integrity of thick metal components such as reactor vessels. Mode I stress intensity factor, denoted by K_I, provides a measure of the intensification of the stress field in a crack tip region. The margin of safety for any given structure is then computed by comparing K_I against the plane strain material toughness.

Several investigators have obtained K_I distributions using 3D finite element techniques. However, they found it necessary to model the crack tip region for each crack position and geometry analyzed. This is both time consuming and expensive.

To accommodate virtually arbitrary three dimensional geometries, a finite element computer program FM3D was developed by Westinghouse Electric Corp. [7].

*Partially supported by USAF Office of Scientific Research Contract No. F44620-76-C-0104.

Badri N. Sahney (ed.), Polynomial and Spline Approximation, 69–81.

Substructuring was used to model the structure so as to isolate a crack region by a single substructure compatible on three faces with standard 20 node brick elements. In essence a crack tip macro-element was developed which eliminates the need for repeated expensive and laborious crack tip modelling. The macro-element was built out of blended [1,4] brick and wedge elements, developed especially for this purpose.

Parks [9] presented a finite element technique for determination of stress intensity factors based on energy release rate. We used this approach in [7] for our three dimensional analyses. Once a finite element solution {u} has been found for an elastic body containing a crack, the potential energy can be written as

$$P = 1/2 \{u\}^T [K] \{u\} - \{u\}^T\{f\} \tag{1}$$

where [K] is the master or global stiffness matrix and {f} is the vector of nodal loads.

A node on the crack line is displaced an amount $\Delta\ell_0$ normal to the crack line. Let $[K]_\ell$ be the stiffness matrix before this node is perturbed and $[K]_{\ell+\Delta\ell_0}$ be the stiffness matrix after perturbation. Note that very few entries of $\Delta[K] = [K]_{\ell+\Delta\ell_0} - [K]_\ell$ are non-zero due to the localness of the finite element method. Similarly, let $\Delta\{f\} = \{f\}_{\ell+\Delta\ell_0} - \{f\}_\ell$ be the perturbation in the load vector due to the displaced node. The incremental change in potential energy is

$$-\Delta P = -1/2 \{u\}^T \Delta[K] \{u\} + \{u\}^T \Delta \{f\} \tag{2}$$

and this change is equal to the weighted average of the energy release rate G(s) as a function of position s along the crack line, that is,

$$-\Delta P = \int_{\text{crackline}} G(s)\, \Delta\ell(s)\, ds. \tag{3}$$

As in [9] we assume that locally

$$G(s) = \frac{1-\nu^2}{E} K_I^2(s). \tag{4}$$

Equating the right-hand sides of (2) and (3) and using (4) we can compute the stress intensity factor K_I associated with the perturbed node.

1. CONSTRUCTION OF MACRO ELEMENT

In this section we describe the geometric description of the macro element itself. Our approach is in the same spirit as that proposed in Cavendish, Gordon and Hall [1,2] in that blended interpolation is used to locally refine the mesh and that a macro or super element is constructed and treated as a substructure during the assembly and solution process.

Our objective is to construct a "brick-like" macro element which (Cf. Figure 1):

1. contains an elliptical crack line in one face,
2. is compatible with the 20-node isoparametric brick on three faces,
3. has variable nodal density in the vicinity of the crack line,
4. permits a wide variety of loadings on the crack face,
5. is parametrically defined so as to allow curved faces (except the face containing the elliptical flaw), and
6. permits variable lengths of the major and minor axes defining the crack line.

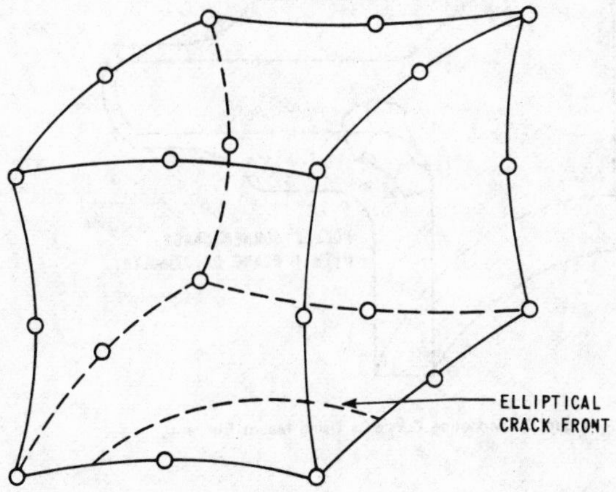

Figure 1 Macro Element

Such a macro element can then be used as a building block to model a variety of semi-elliptical cracks (Cf. Figure 2) or at the extreme, eight such macro elements can be used to model an elliptical pocket. In Figure 2, note that 20-node elements can be adjoined to the macro elements on all faces except the face containing the crack.

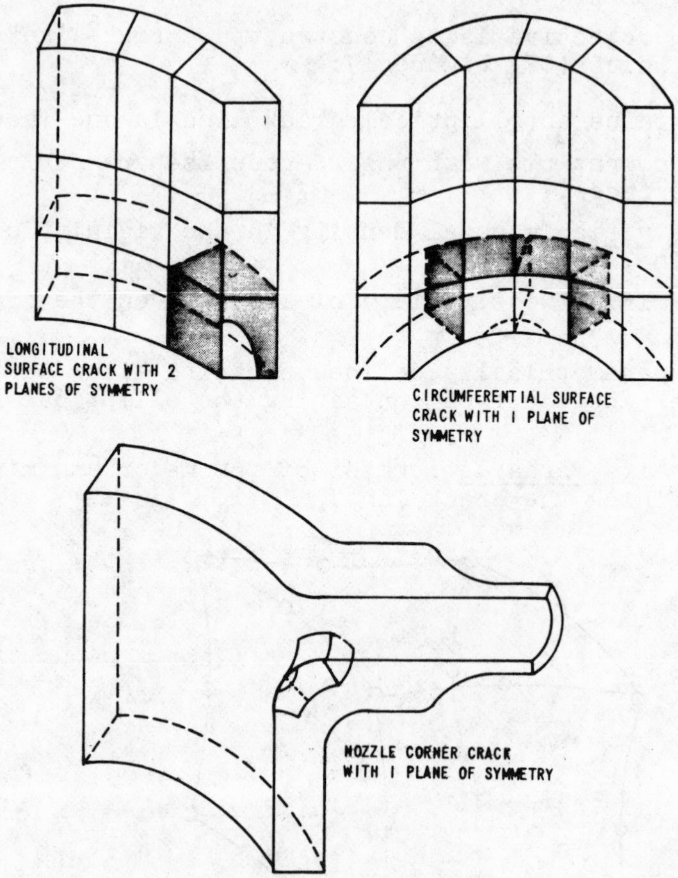

Figure 2 Modeling Concepts Using Macro Element

The undeformed macro element constructed for this investigation is illustrated in Figure 3 and consists of a channel and a transition region. The channel surrounds the crack line region with a high density of nodes while the transition region fills the gap between the channel and the faces of the macro element. The channel consists of 28 micro elements or cells and the transition region consists of 17 micro elements for a total of 45 micro elements in the macro element.

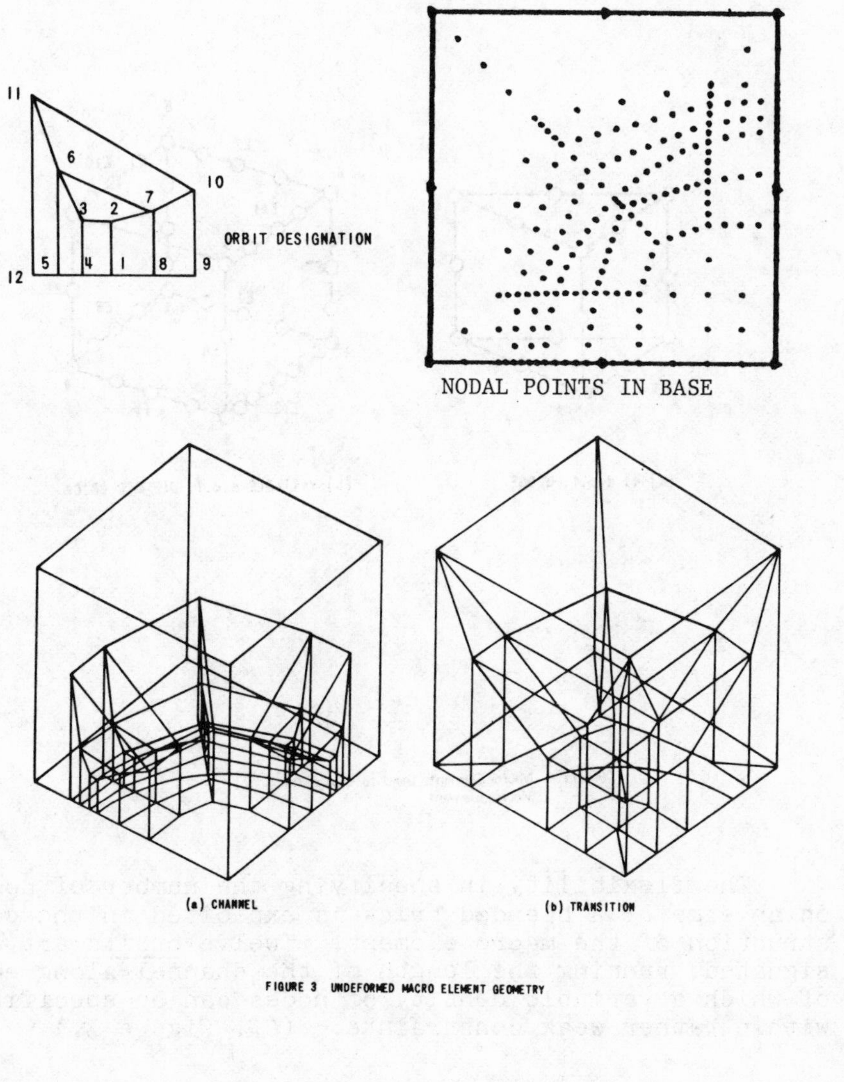

FIGURE 3 UNDEFORMED MACRO ELEMENT GEOMETRY

The micro elements are of two types: a 15-node wedge [10] and a variable-node blended brick element [6]. The wedge has 3 nodes along each of its 9 edges and is a 45 degree of freedom element with quadratic shape functions. (Cf. Figure 4a).

The blended brick element (Cf. Figure 4b) is defined in detail in [6]. See also [2]. Here it suffices to say that the shape functions are piecewise quadratics and the user specifies twelve integers N1,..., N12 which determine the number of nodes along each edge.

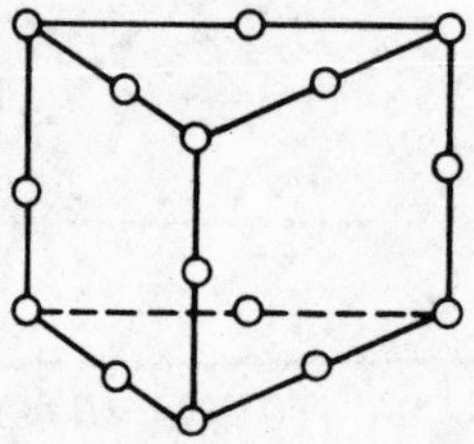

(a) 45 d.o.f. WEDGE

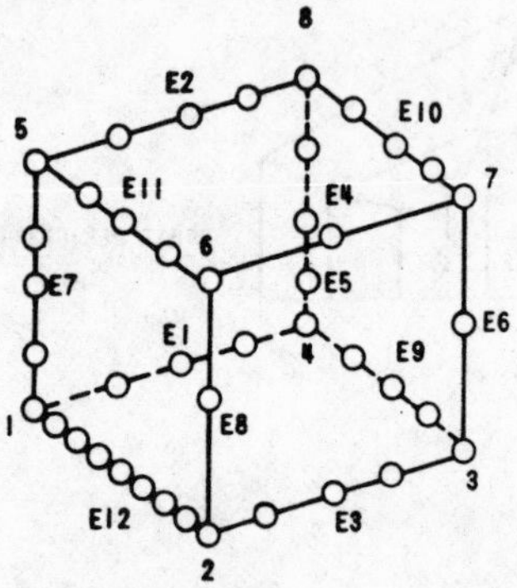

(b) VARIABLE d.o.f. BLENDED BRICK

Figure 4 Micro Elements Used to Design the Macro Element

The flexibility in specifying the number of nodes on an edge of a blended brick is exploited in the construction of the macro element. Twelve orbits are designated, running the length of the channel along each of which a variable density of nodes can be specified within rather weak constraints. (Cf. Figure 3.)

To be useful in modelling surface flaws in 3-D structures with asymmetric surfaces, the macroelement must allow for curved faces. This is accomplished by constructing a vector valued function or domain transformation $\bar{T}$ from the unit cube $\mathcal{S}$: [0,1] × [0,1] × [0,1] onto the desired three dimensional element $\mathcal{E}$ in such a way that the circular crack line is mapped onto an elliptical crack line in $\mathcal{E}$, Figure 5.

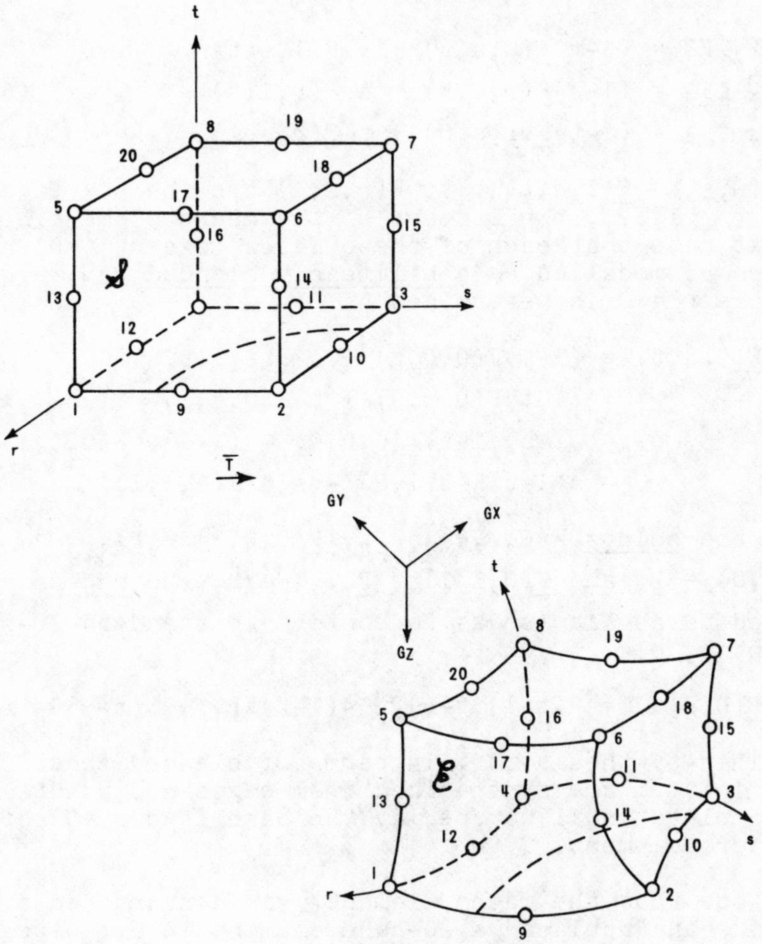

Figure 5. Parametric Transformation.

The tri-linearly blended formula [3] is used to construct $\bar{T}$ which provides a relation between the global coordinates (GX,GY,GZ) and the generalized (r,s,t) coordinates of a point in $\mathcal{E}$:

$$\bar{T}(r,s,t) = \begin{pmatrix} GX(r,s,t) \\ GY(r,s,t) \\ GZ(r,s,t) \end{pmatrix} = \mathcal{P}_r \oplus \mathcal{P}_s \oplus \mathcal{P}_t [\underline{F}] \tag{5}$$

$$= (\mathcal{P}_r + \mathcal{P}_s + \mathcal{P}_t - \mathcal{P}_r\mathcal{P}_s - \mathcal{P}_r\mathcal{P}_t - \mathcal{P}_s\mathcal{P}_t + \mathcal{P}_r\mathcal{P}_s\mathcal{P}_t)[\underline{F}]$$

where

$$\begin{aligned} \mathcal{P}_r[\underline{F}] &= (1-r)\underline{F}(0,s,t) + r\,\underline{F}(1,s,t) \\ \mathcal{P}_s[\underline{F}] &= (1-s)\underline{F}(r,0,t) + s\,\underline{F}(r,1,t) \\ \mathcal{P}_t[\underline{F}] &= (1-t)\underline{F}(r,s,0) + t\,\underline{F}(r,s,1) \end{aligned} \tag{6}$$

and $\underline{F}(0,s,t)$, $\underline{F}(1,s,t)$, ..., $\underline{F}(r,s,1)$ are the six faces $r = 0$, $r = 1$, ..., $t = 1$ resp. of the macro element $\mathcal{E}$. We next note that each of these faces (except t=0) can in turn be modelled by a bilinearly blended map, e.g. the face r = 0 is determined by

$$\begin{aligned} \underline{F}(0,s,t) = &(1-s)\underline{F}(0,0,t) + s\,\underline{F}(0,1,t) \\ &+ (1-t)\underline{F}(0,s,0) + t\,\underline{F}(0,s,1) \\ &- (1-s)(1-t)\underline{F}(0,0,0) - (1-s)t\,\underline{F}(0,0,1) \\ &- s(1-t)\underline{F}(0,1,0) - s\,t\,\underline{F}(0,1,1) \end{aligned} \tag{7}$$

where the points $\underline{F}(0,0,0) = \underline{P}_4$, $\underline{F}(0,0,1) = \underline{P}_8$, $\underline{F}(0,1,0) = \underline{P}_3$ and $\underline{F}(0,1,1) = \underline{P}_7$. Next, the edges are assumed parabolic (so as to conform to standard 20-node elements), e.g.

$$\underline{F}(0,0,t) = 2(t-1)(t-\tfrac{1}{2})\underline{P}_4 + 4(1-t)t\underline{P}_{16} + 2t(t-\tfrac{1}{2})\underline{P}_8. \tag{8}$$

This synthesis of $\bar{T}$ in terms of blended faces which in turn are synthesized from edges and points is not complete until we specify the base (face t=0) of the macro element.

Note that the macro element $\mathcal{E}$ is constructed so that it can "replace" a 20-node element in a finite element idealization of a 3-D structure. Hence, a structure can be modelled once and for all as a catenation of 20 node isoparametric elements. The location of an elliptical flaw in this idealization can then be varied by replacing 1, 2, 4 or 8 20-node elements by the appropriate combination of macro elements. Interfacial compatibility is guaranteed. Cf. Figure 2.

The mapping $\bar{T}$ in (5) has the property that the 20-nodes of $\mathcal{S}$ are mapped into the respective 20 nodes of $\mathcal{E}$. Further, faces are mapped onto respective faces. The image of all but one of these faces (t=0) has been chosen to be the same as if $\bar{T}$ were the standard isoparametric mapping [10]. Choosing $\underline{F}(r,s,0)$ as the appropriate linearly blended interpolant of the four boundary edges would, in fact, reduce $\bar{T}$ to the standard 20-node isoparametric mapping. For such a choice, however, the circular crack line in $\mathcal{S}$ would, in general, be grossly distorted in the base of $\mathcal{E}$. This situation is remedied by choosing the image of the face t = 0 as follows. We assume the crack line in $\mathcal{E}$ is given as a segment of an ellipse with center at node 2 and axes of length 2CXA and 2CYB. The base of $\mathcal{E}$ is assumed planar and the (CX,CY,CZ) crack coordinate system is given in terms of rotation angles $(\alpha_1,\alpha_2,\alpha_3)$ relative to the global coordinate system.

The crack coordinate system coordinates (CX21,CY21,CZ21) of point 21 (and 22) are determined by the method of bisection applied to the two equations describing the ellipse and the parabola through that point, Figure 6.

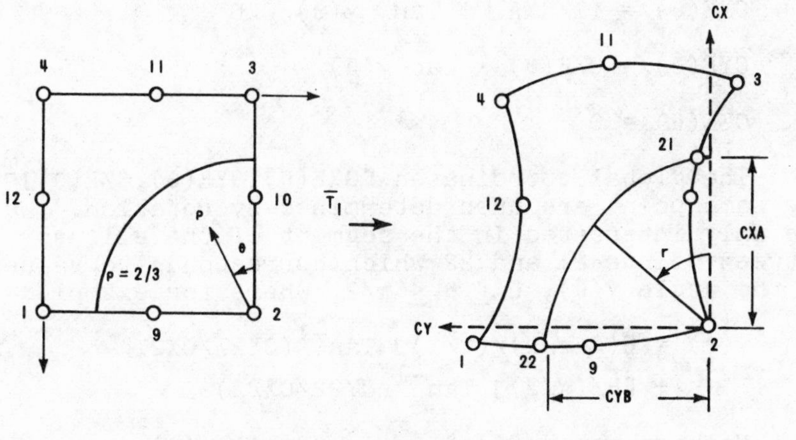

Fig. 6. Crack Face Transformation.

We next reference points in the unit square by the (ρ,θ) system so that $\rho = 2/3$ is the crack line in the undistorted or parameter domain. Such a mapping is derivable for example using <u>linear blending in θ</u> and <u>quadratic blending in ρ</u> as

$$\bar{T}_1(\rho,\theta) = \begin{pmatrix} GX(\rho,\theta) \\ GY(\rho,\theta) \\ GZ(\rho,\theta) \end{pmatrix} = B_2(\rho) \begin{pmatrix} GXE(\theta) \\ GYE(\theta) \\ GZE(\theta) \end{pmatrix} + B_1(\rho) \begin{pmatrix} X_2 \\ Y_2 \\ Z_2 \end{pmatrix} + B_3(\rho) \begin{pmatrix} GXD(\theta) \\ GYD(\theta) \\ GZD(\theta) \end{pmatrix} \quad (9)$$

where

$$\begin{aligned} B_1(\rho) &= 3(\rho-2/3)(\rho-\rho_1(\theta))/(2\rho_1(\theta)), \\ B_2(\rho) &= 3\rho(\rho-\rho_1(\theta))/(2(2/3-\rho_1(\theta)), \\ B_3(\rho) &= \rho(\rho-2/3)/(\rho_1(\theta)(\rho_1(\theta)-2/3)), \end{aligned} \quad (10)$$

are the quadratic blending functions, and

$$\rho_1(\theta) = \begin{cases} \text{SEC } \theta, & 0 \leq \theta \leq \pi/4 , \\ \text{CSC } \theta, & \pi/4 \leq \theta \leq \pi/2 . \end{cases} \quad (11)$$

The crack coordinates of a point on the ellipse are determined by

$$\begin{aligned} CXE(\theta) &= \{1/CXA^2 + \text{Tan}^2\, \gamma(\theta)/CXB^2\}^{-1/2} , \\ CYE(\theta) &= CXE(\theta) \cdot \text{Tan}\, \gamma(\theta) , \\ CZE(\theta) &= 0 . \end{aligned} \quad (12)$$

The global coordinates $[GXE(\theta),GYE(\theta),GZE(\theta)]^T$ of the same point are then determined by rotation. We are only interested in the segment of the ellipse between points 21 and 22 which corresponds to values of the angle $\gamma(\theta)$, $0 \leq \theta \leq \pi/2$, where for example

$$\begin{aligned} \gamma(\theta) = &[(\theta-\pi/2)/(\pi/2)]\, \text{Tan}^{-1}(CY21/CX21) \\ &+ [\theta/(\pi/2)]\, \text{Tan}^{-1}(CY22/CX22) . \end{aligned} \quad (13)$$

We have now defined every term in$_T$(9) except the parametrization $[GXD(\theta),GYD(\theta),GZD(\theta)]^T$ of the edges E9 and E1 of $\mathcal{E}$, which we now specify as:

$$\begin{pmatrix} GXD(\theta) \\ GYD(\theta) \\ GZD(\theta) \end{pmatrix} = 4(\theta-\eta)(\theta-\pi/4)/(\pi\eta)\bar{P}_3+(\pi/4-\theta)\theta/(\eta(\pi/4-\eta))\bar{P}_{11}$$
$$+ 4\theta(\theta-\eta)/(\pi(\pi/4-\eta))\bar{P}_4, \quad 0 \leq \theta \leq \pi/4$$
$$= 4(\theta-\pi/2)(\theta-\mu)/(\pi(\mu-\pi/4))\bar{P}_4 \qquad (14)$$
$$+ (\pi/2-\theta)(\theta-\pi/4)/(\pi/2-\mu)(\mu-\pi/4)\bar{P}_{12}$$
$$+ 4(\theta-\mu)(\theta-\pi/4)/(\pi(\pi/2-\mu))\bar{P}_1, \quad \pi/4 \leq \theta \leq \pi/2,$$

where $\eta = \mathrm{Tan}^{-1}(.5)$ and $\mu = \mathrm{Tan}^{-1}(2)$. As θ varies from 0 to $\pi/2$, the point $(GXD(\theta),GYD(\theta),GZD(\theta))$ traverses the path along edge 9 from point 3 to point 4, then along edge 1 to point 1. The map $\bar{T}_1$ treats the crack face somewhat independent of the region bounded by the crack line and edges 1 and 9.

Figure 7A-B illustrate macro elements which are images of a unit cube under the blended vector valued mapping (5). Results of using such macro elements in computing stress intensity factors are reported in [7,11].

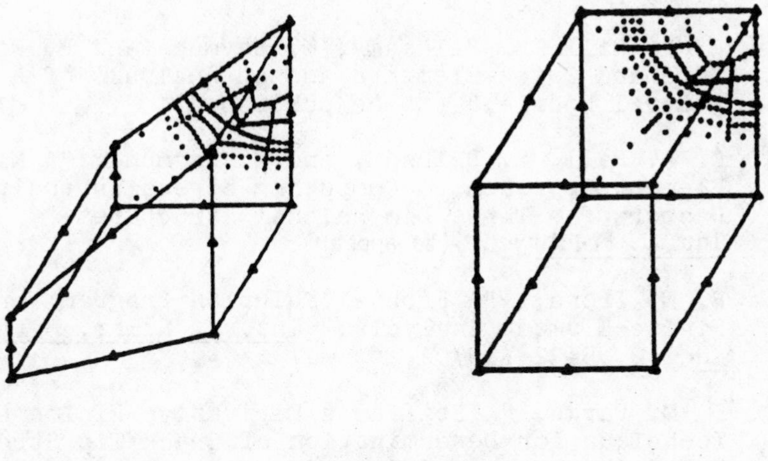

Figure 7A. Macro elements with crack face detail.

REFERENCES

1. J. C. Cavendish, W. J. Gordon and C. A. Hall, "Ritz-Galerkin Approximations in Blending Function Spaces," Numer. Math. 26, 155-178 (1976).

2. J. C. Cavendish, W. J. Gordon and C. A. Hall, "Substructured Macroelements Based on Locally Blended Interpolation," Int. J. Numer. Mech. Engg. 11, 1405-1421 (1977).

3. W. J. Gordon, "Distributive Lattices and the Approximation of Multivariate Functions," in Approximations with Special Emphasis on Spline Functions: Proceedings, 1969, editor I. J. Schoenberg, Academic Press, New York, 1969, pp. 223-277.

4. W. J. Gordon and C. A. Hall, "Transfinite Element Methods: Blending-Function Interpolation Over Arbitrary Curved Element Domains," Numer. Math. 21, 109-129 (1973).

5. C. A. Hall, "Transfinite Interpolation and Applications to Engineering Problems," in Theory of Approximation: With Applications, editors A. G. Law and B. N. Sahney, Academic Press, New York, 1976, pp. 308-331.

6. C. A. Hall, S. Palusamy, M. Raymund and F. Wimberley, "Blended Brick Elements and an Analysis of a Lateral Nozzle," (to be published).

7. C. A. Hall, S. Palusamy and M. Raymund, "A Macro Element Approach to Computing Stress Intensity Factors for Three Dimensional Structures," Int. J. Fracture, (to appear).

8. B. M. Irons, "A Frontal Solution Program for Finite Element Analysis," Int. J. Numer. Methods Engg 2, 5-32 (1970).

9. D. M. Parks, "A Stiffness Derivative Finite Element Technique for Determination of Crack Tip Stress Intensity Factors," Int. J. Fracture 10, 487-502 (1974).

10. O. C. Zienkiewicz, The Finite Element Method in Engineering Science, McGraw-Hill, New York, 1971.

11. J. J. McGowan and M. Raymund, "Stress Intensity Factor Solutions for Internal Longitudinal Semi-Elliptical Surface Flaws in a Cylinder Under Arbitrary Loadings," presented at the Eleventh National Symposium on Fracture Mechanics, June 12-14, 1978, Blacksburg, VA.

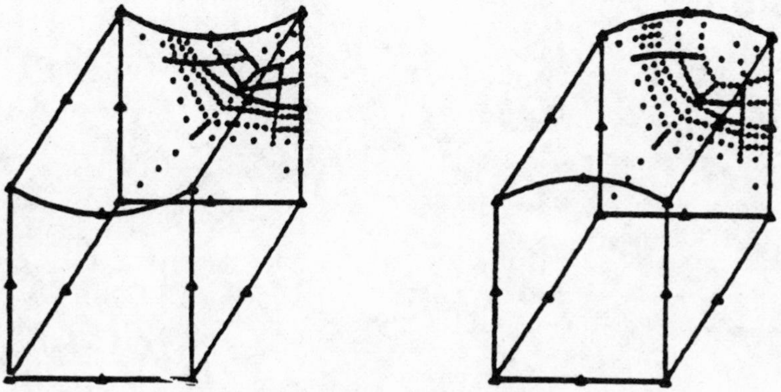

Figure 7B. Macro elements with crack face detail.

SIMULTANEOUS APPROXIMATION OF FUNCTION AND DERIVATIVE ON $[0,\infty]$ AND AN APPLICATION TO INITIAL VALUE PROBLEMS

Charles Hall*
Institute for Computational Mathematics and Applications
Department of Mathematics and Statistics
University of Pittsburgh

ABSTRACT. Convergence of the approximation of a class of functions defined on $[0,\infty]$ and their derivatives by exponomials and Laguerre functions is established. These results are used in the analysis of a new finite element type for initial value problems.

INT. Let $f(x)$ be defined on $[0,1]$; then the nth Berstein polynomial associated with f is given by

$$B_n(f,x) = \sum_{k=0}^{n} f(k/n)\binom{n}{k} x^k(1-x)^{n-k} \quad . \tag{1}$$

The classic result of Berstein [5, Theorem 6.2.2] which establishes the Weierstrass Polynomial Approximation Theorem is that if $f \in C[0,1]$ then

$$\lim_{n\to\infty} B_n(f,x) = f(x) \quad \text{uniformly in } [0,1]. \tag{2}$$

In contrast to many other approximants, the Berstein polynomials also yield good approximations to derivatives of f (when they exist). In fact, [5, Theorem 6.3.2] if $f \in C^p[0,1]$ then

$$\lim_{n\to\infty} B_n^{(p)}(f,x) = f^{(p)}(x) \quad \text{uniformly on } [0,1]. \tag{3}$$

Recently, the analysis of a new finite element type [3,4] precipitated the need for the extension of this notion of simultaneous approximation of a function and its derivative for functions defined on the semi-infinite interval $[0,\infty]$.

---

*Partially supported by USAF Office of Scientific Research Contract No. F44620-76-0104.

Badri N. Sahney (ed.), Polynomial and Spline Approximation, 83–90.

1. Infinite Elements.

Consider the initial-boundary value problem

$$\frac{\partial u}{\partial t} = L[u] + f(u) \qquad 0 < x < 1 \quad t > 0 , \tag{4}$$

$$\begin{aligned} \alpha_0 \frac{\partial u}{\partial x}(0,t) + \beta_0 \, u(0,t) &= \delta_0 , \\ \alpha_1 \frac{\partial u}{\partial x}(1,t) + \beta_1 \, u(1,t) &= \delta_1 , \end{aligned} \tag{5}$$

$$u(x,0) = g(x) \tag{6}$$

where L is an elliptic spatial differential operator and α_i, β_i and δ_i are constants. We assume that the solution at $t = \infty$ is known or has been found numerically. In [3,4] a new method is presented which for such problems produces an approximation to u for all (x,t) upon the solution of a single system of algebraic equations. The independent variable domain $S = \{(x,t) | 0 \leq x \leq 1, t \geq 0\}$ is partitioned into N+1 semi-infinite strips or infinite elements $[x_{i-1}, x_i] \times [0,\infty)$ where $0 = x_0 < x_1 < \ldots < x_{N+1} = 1$. In each infinite element the solution to (4)-(6) is approximated by a fully discretized finite linear combination of shape functions and an associated stiffness matrix is computed using Galerkin's method. The assembly and solution proceeds as if the original problem were void of a transient term and the initial and steady state conditions are treated as essential boundary conditions.

In the notation of Galerkin's method u is approximated by

$$U(x,t) = \sum_{i=0}^{N+1} \sum_{j=0}^{M} a_{ij} \, \theta_j(t) \, \phi_i(x) \tag{6}$$

where $\{\phi_i(x)\}_{i=0}^{N+1}$ are continuous linear splines on the mesh $0 = x_0 < x_1 < \ldots < x_{N+1} = 1$ such that

$$\phi_i(x_j) = \delta_{ij}, \quad 0 \leq i,j \leq N+1 ,$$

and for the basis functions $\theta_j(t)$ we consider two choices (Cf. [3,4])

(i) Exponomials: $\theta_0(t) = 1,$
$\theta_1(t) = e^{-\alpha t}, \quad \alpha > 0,$
$\theta_n(t) = e^{-n\alpha t} - e^{-\alpha t}, \quad n = 2,3,\ldots, M,$

(ii) Laguerre Functions: $\theta_0(t) = 1,$
$\theta_1(t) = e^{-\alpha t}, \quad \alpha > 0,$
$\theta_n(t) = t^{n-1}e^{-\alpha t}, \quad n=2,3,\ldots M.$

We note that from given data, $a_{i0} = u(x_i,\infty)$ and $a_{i1} = g(x_i) - u(x_i,\infty)$. Certain of the coefficients a_{0j} and $a_{N+1,j}$ may be determined *a priori* by the boundary conditions also. The other coefficients are determined from the Galerkin equations

$$\int_0^1 \int_0^\infty \left(\frac{\partial U}{\partial t} - L[U] - f(U)\right)\phi_k(x)\, \theta_\ell(t)\,dt\,dx = 0, \quad (7)$$

$0 \leq k \leq N+1, \ 2 \leq \ell \leq M.$

This numerical scheme, first introduced in [3] was analyzed in [4] as two applications of Galerkin's method. That is, the system of equations in (7) is "identical" to the system of equations one would obtain if the following procedure was followed:

First, the semi-discrete Galerkin approximation [8]

$$\hat{u}(x,t) = \sum_{i=0}^{N+1} \hat{u}_i(t)\phi_i(x) \quad (8)$$

is formed as the solution to the system of ordinary differential equations

$$\int_0^1 \left(\frac{\partial \hat{u}}{\partial t} - L[\hat{u}] - f(\hat{u})\right)\phi_i(x)\,dx = 0, \quad (9)$$

$1 \leq i \leq N.$ (Assuming Dirichlet boundary conditions)

Second, the system (9) which we now write generically as

$$dy/dt = -F(t,y), \quad 0 < t < b \leq \infty, \quad (10)$$

where y is an N-vector of continuously differentiable functions (In our case $y(t) = (\hat{u}_1, \hat{u}_2, \ldots, \hat{u}_N)$), is solved by Galerkin's method. This time y(t) is approximated by

$$Y(t) = y(b) + \sum_{n=1}^{M} c_n \theta_n(t), \quad c_n \in R^N, \tag{11}$$

(In our case $b = \infty$.) We need require, as before, $\theta_1(0) = 1$, $\theta_n(0) = 0$, $n = 2,3,\ldots, M$, and $\lim_{t\to b} \theta_n(t) \equiv \theta_n(b) = 0$, $1 \le n \le M$. Furthermore, $c_1 = y_0 - y(b)$ and the remaining vector coefficients c_n, $n = 2,\ldots, M$, are determined by

$$\int_0^b \Big[dY/dt + F(t,Y)\Big]_i \theta_j(t)dt = 0 , \tag{12}$$

$1 \le i \le N$, $2 \le j \le M$. (Again $b = \infty$ in our case.)

The advantage of viewing (7) as a composite of (9) and (12) is that it is well known [6,8] that under suitable conditions

$$||\hat{u}-u||_T \to 0 \quad \text{as } N \to \infty \tag{13}$$

where $||\cdot||_T = \int_0^T \int_0^1 |\hat{u}-u|^2 dxdt$. Hence in order to establish convergence of $U \to u$, one need only show

$$Y(t) \to y(t) = (\hat{u}_1, \hat{u}_2, \ldots, \hat{u}_N)^T \tag{14}$$

as $M \to \infty$. This is precisely the result of the following Theorem of J. Cavendish, T. Porsching and the author:

THEOREM 1 (Cavendish, Hall, Porsching [4]). *Let $F: (0,b) \times R^N \to R^N$ be a continuously differentiable, symmetric, uniformly monotone map on R^N which is also uniformly Lipschitz in t. Let $D \subset R^N$ be any compact set which contains $y(t)$, $t \in [0,b]$ where $y(t)$ is the solution to (10). If W is any element of*

$$S_M = \{w| \ w=y(b)+ \sum_{n=1}^{M} a_n\theta_n(t), \ a_1=y_0-y(b)\} \tag{15}$$

which is also in D for $0 \le t \le b$, and if Y is the Galerkin approximation determined by (12), then

$$||Y-y||_2 \le \alpha_1 \ ||W-y||_2 + \alpha_2 \ ||d(W-y)/dt||_2, \tag{16}$$

where the constants α_1 and α_2 do not depend on W or Y.

Hence we can prove that $Y \to y$ or equivalently $U \to u$ *if* we can establish that y and dy/dt can be simultaneously approximated by <u>some</u> function $W \varepsilon S$ and its derivative respectively.

There are boundary control problems ([4,7] and references cited therein) for which the value of u in (4) is specified at $t = b < \infty$ and the problem is to find the appropriate boundary data in (5) of a specific form. For such problems ($b<\infty$) there are many ways to choose the expansion set $\{\theta_n(t)\}_{n=1}^{M}$ such that the right side of (16) is arbitrarily small for M sufficiently large. For example, continuous linear splines on the mesh $0 = t_0 < t_1 <\ldots< t_{M+1} = b$ would yield in (16) using standard error bounds [1]

$$||Y-y||_2 = O(1/M) \quad \text{as } M \to \infty, \tag{17}$$

under the assumption that $y \varepsilon C^2(0,b)$. As the regularity of y improves, the choice of higher order piecewise polynomials for θ_n yields more rapid convergence of the bounds in (16).

Another choice for the case $b < \infty$ is $\theta_n(t)$ to be a polynomial of degree n. As we already recalled, there exist polynomials, e.g. Bernstein polynomials which simultaneously approximate y and dy/dt as closely as desired.

Contrasted to this, the case $b = \infty$ is less routine and the results of [4] are summarized in the next two sections.

2. <u>Exponomials</u>

<u>THEOREM 2</u> (Cavendish, Hall, Porsching [4]). <u>Let</u> $x \varepsilon C^1(0,\infty)$ <u>and suppose</u> x <u>and</u> dx/dt <u>vanish at</u> ∞ <u>and belong to</u> $L_2(0,\infty)$. <u>If</u> $\epsilon > 0$, <u>then there is an exponomial</u> h(t) <u>such that</u> h(0) = x(0) <u>and simultaneously</u>

$$||x-h||_2 \leq \epsilon \quad \text{and} \quad ||d(x-h)/dt||_2 \leq \epsilon. \tag{18}$$

PROOF: The proof given in [4] makes use of the following three Lemmas.

<u>Lemma 1</u>. <u>There is a</u> $g \varepsilon C^1(0,\infty)$ <u>and</u> a t_0 <u>such that</u> g(0) = x(0), $g \equiv 0$ <u>for</u> $t > t_0$ <u>and</u>

$$||x-g||_2 \leq \epsilon \quad \text{and} \quad ||d(x-g)/dt||_2 \leq \epsilon. \qquad (19)$$

Lemma 2. There is an exponomial $h_1(t)$ such that $h_1(0) = g(0)$ and

$$||e^{\alpha t} g-h_1||_\infty \leq \epsilon/\alpha \text{ and } ||e^{\alpha t} d(e^{\alpha t} g-h_1)/dt||_\infty \leq \epsilon. \qquad (20)$$

Lemma 2 is used to prove,

Lemma 3. There is an exponomial $h(t)$ such that $h(0) = g(0)$ and

$$||g-h||_2 \leq \epsilon/\sqrt{2}\alpha \text{ and } ||d(g-h)/dt||_2 \leq (1+\alpha)\epsilon/\sqrt{2}\alpha. \qquad (21)$$

(Hint: Let $h(t) = e^{-\alpha t} h_1(t)$.)

Theorem 2 can then be used to establish the convergence of the exponomial Galerkin approximation U to u or equivalently Y to y. To wit,

THEOREM 3 (Cavendish, Hall, Porsching [4]). Let $y(t)$ be the unique solution to (10) on $[0,\infty)$ and let $y(\infty)$ be its steady state value. Let $Y(t)$ be the exponomial Galerkin approximation in (11). If $y - y(\infty)$ and dy/dt belong to $L_2(0,\infty)$ then under the hypotheses of Theorem 1,

$$\int_0^\infty |Y-y|^2 dt \to 0 \quad \text{as } M \to \infty. \qquad (22)$$

Proof: Let $x(t) = y(t) - y(\infty)$. By Theorem 2 there is an exponomial $h(t) = \sum_{n=1}^{M} c_n e^{-n\alpha t}$ such that $y(0) - y(\infty) = \sum_{n=1}^{M} c_n = h(0)$, and such that

$$||x-h||_2 < \epsilon \quad \text{and} \quad ||d(x-h)/dt||_2 < \epsilon.$$

Let $W = y(\infty) + (y_0-y(\infty))\theta_1(t) + \sum_{n=2}^{M} c_n \theta_n$; then

$$||y-W||_2 = ||x-h||_2 < \epsilon$$

and similarly for the first derivatives.

The result follows from substituting this particular W into (16). Q.E.D.

3. Laguerre Functions.

According to [9, Theorem 18], any continuous function which vanishes at ∞ can be uniformly approximated by a Laguerre function. This result was used to prove

THEOREM 4 (Cavendish, Hall, Porsching [4]). *Let* $x \in C^1(0,\infty)$ *and suppose* x *and* dx/dt *vanish at* ∞ *and belong to* $L_2(0,\infty)$. *If* $\epsilon > 0$ *then there is a Laguerre function* $p(t)e^{-\alpha t}$ *such that* $p(0) = x(0)$, *and*

$$||x-pe^{-\alpha t}||_2 \leq \epsilon \text{ and } ||d(x-pe^{-\alpha t})/dt||_2 \leq \epsilon. \quad (23)$$

Proof: The proof given in [4] makes use of the Lemma 1 and the following two lemmas.

Lemma 4. *There is a Laguerre function* $p(t)e^{-2\alpha t/3}$ *such that* $p(0) = g(0)$ *and*

$$||e^{\alpha t/3}g-pe^{2\alpha t/3}||_\infty \leq 3\epsilon/\alpha,$$
$$||e^{\alpha t/3}d(e^{\alpha t/3}g-pe^{-2\alpha t/3})/dt||_\infty \leq \epsilon. \quad (24)$$

The L_2 estimates then follow in

Lemma 5. *There is a Laguerre function* $p(t)e^{-\alpha t}$ *such that* $p(0) = g(0)$ *and*

$$||g-pe^{-\alpha t}||_2 \leq \epsilon(3/2\alpha)^{\frac{1}{2}},$$
$$||d(g-pe^{-\alpha t})/dt||_2 \leq \epsilon(1+\alpha/3)(3/2\alpha)^{\frac{1}{2}}.$$

(Hint: $g - pe^{-\alpha t} = e^{-\alpha t/3}(e^{\alpha t/3}g-pe^{-2\alpha t/3})$.)

Theorem 4 can be used to establish the convergence of the Laguerre Galerkin approximation U to u, or equivalently Y to y.

For several examples of the computer implementation of the infinite element method to parabolic problems, see [2,3,4].

REFERENCES

1. G. Birkhoff, M. Schultz and R. S. Varga, "Piecewise Hermite Interpolation in One and Two Variables with Applications to Partial Differential Equations," Numer. Math. 11, (1968), 232-256.

2. J. C. Cavendish and C. A. Hall, "Blended Infinite Elements for Parabolic Boundary Value Problems," General Motors Research Publication GMR-121, 1977.

3. J. C. Cavendish, C. A. Hall and O. C. Zienkiewicz, "Blended Infinite Elements for Parabolic Boundary Value Problems," Int. J. Num. Meth. Engg. (to appear).

4. J. C. Cavendish, C. A. Hall and T. A. Porsching, "Galerkin Approximations for Initial Value Problems with Known End Time Conditions," submitted for publication.

5. P. J. Davis, Interpolation and Approximation, Blaisdell, London, 1963.

6. J. Douglas, Jr. and T. Dupont, "Galerkin Methods for Parabolic Equations," SIAM J. Numer. Anal. 7, (1970), 575-626.

7. P. J. Harley and A. R. Mitchell, "A Finite Element Collocation Method for the Exact Control of a Parabolic Problem," Int. J. Num. Meth. Engg. 11, (1977), 345-353.

8. H. S. Price and R. S. Varga, "Error Bounds for Semidiscrete Galerkin Approximations of Parabolic Problems with Applications to Petroleum Reservoir Mechanics," in Numerical Solution of Field Problems in Continuum Physics, SIAM-AMS Proc. II, AMS, 1970.

9. M. H. Stone, "A Generalized Weierstrass Approximation Theorem," in Studies in Modern Analysis, 1, R. C. Buck, ed., MAA-Prentice-Hall, Englewood Cliffs, NJ, 1962.

COMPOSITE METHODS FOR GENERATING SURFACES

Peter Lancaster
The University of Calgary

ABSTRACT

A central theme concerning interpolating sets and precision classes of linear transformations is developed. The ideas are illustrated with two sequences of topical examples:
(a) Smooth surface construction on scattered data and
(b) The numerical implementation of blending on rectangular grids. Special attention is drawn to composite methods involving pairs of projectors which do not commute.

INTRODUCTION

This contribution consists of some general remarks, mostly of an algebraic character, on the composition of interpolating projectors and on their Boolean sums. The ideas are illustrated by two sequences of examples each sequence being important in its own right. These concern (a) surface construction on scattered data and (b) the numerical implementation of blending on rectangular grids. Special attention is drawn to pairs of projectors P,Q for which $QP \neq PQ$, but QP is a projector.

No attempt is made to compete with recent comprehensive reviews by Barnhill [1], Lancaster and Salkauskas [8], Powell [10], and Schumaker [13].

PRELIMINARY IDEAS AND EXAMPLES

Let X be a real linear space of continuous functions all having common domain $\Omega \subset \mathbb{R}^2$ and $L(X)$ the algebra of linear transformations on X. A transformation $T \in L(X)$ is said to have *interpolating set* $\Gamma_T \subset \Omega$ if $Tf = f$ on Γ_T for any $f \in X$. Transformation T has *precision class* $S_T \subset X$ if $Tf = f$ for all $f \in S_T$. In other words, the precision class of T is Ker$(I-T)$.

If $P \in L(X)$ and $P^2 = P$, then P is called a *projector* and Ker$(I-P)$ = Im P gives another description of the precision class. A projector P is called an *interpolating projector* if its interpolating set is not empty.

Badri N. Sahney (ed.), Polynomial and Spline Approximation, 91-102.

The composite methods of the title of this paper consist of functional composition and Boolean addition of interpolating projectors, say P and Q. Thus, we are to study the properties of QP and

$$Q \oplus P = P + Q - QP.$$

In many applications interpolating projectors P are such that Im P is finite dimensional and, in such cases, knowledge of the interpolating set Γ_P will sometimes define P completely.

EXAMPLE 1.1 Let $\varphi_1, \ldots, \varphi_M$ be linearly independent linear functionals on $C^{\mu}[a,b]$. In applications they are usually point functionals specifying function or derivative values at points of $[a,b]$. Let S be an M-dimensional subspace of $C^{\mu}[a,b]$ with basis $\hat{g}_1, \ldots, \hat{g}_M$ having the property

$$\text{(1)} \qquad \det \left[\varphi_i(\hat{g}_j)\right]_{i,j=1}^{M} \neq 0 \, .$$

Then the following results are well-known:

LEMMA *Given the condition* (1), *there is a basis* $g_1, \ldots, g_M$ *for* S *such that*

$$\varphi_i(g_j) = \delta_{ij} \, , \qquad 1 \leq i,j \leq M \, .$$

Writing $S^* = \text{span}\,\{\varphi_1, \ldots, \varphi_M\}$, the two sets $\{g_i\}$, $\{\varphi_i\}$ are known as *dual bases* for S and S^*. The first set is also known as a basis of *cardinal functions*.

THEOREM 1 (a) *The transformation defined by*

$$\text{(2)} \qquad P = \sum_{i=1}^{M} g_i(x)\varphi_i$$

is a projection of $C^{\mu}[a,b]$ *onto* S.

(b) *If* $f \in C^{\mu}[a,b]$, *there is a unique* $u \in S$ *such that*

$$\text{(3)} \qquad \varphi_i(u) = \varphi_i(f)\, , \qquad i = 1,2, \ldots, M$$

and $u = Pf$.

As special cases of this result consider, for example, $S = P^{n-1}$, the space of polynomials with degree not exceeding n-1, and having dimension n. If functionals φ_i are defined via n distinct points $x_1, \ldots, x_n$ in $[a,b]$ by

$$\varphi_i(f) = f(x_i), \qquad i = 1,2, \ldots, n \, ,$$

then the functions $g_1, \ldots, g_n$ of the lemma are the fundamental

Lagrange interpolating polynomials. Then

$$\Gamma_P = \{x_1, x_2, \ldots, x_n\} \quad \text{and } S_P = S = P^{n-1} .$$

Alternatively, choose S to be the space of natural cubic splines with knots $a = x_1 < x_2 < \ldots < x_n = b$. This is known to have dimension n and the lemma is known to apply. Again, $\Gamma_P = \{x_1, \ldots, x_n\}$ and $S_P = S$. In both cases the cardinal functions satisfy

$$g_i(x_j) = \delta_{ij} , \qquad i,j = 1,2,\ldots, n.$$

EXAMPLE 2.1 The second sequence of examples concerns the problem of constructing a C^1 surface on the domain Ω which is to interpolate given function values at N randomly placed points in Ω. The abbreviation z will be used for coordinate pairs x,y determining points in $\mathbb{R}^2$. Thus, $z_1, \ldots, z_N$ will denote the N distinct data points at which values $f_1, \ldots, f_N$ are prescribed. For convenience, we take X to be the space of C^1 functions. Surfaces of the required type can be constructed by first applying a moving least squares method (as described in my other paper of this volume). If the data $f_1, \ldots, f_N$ are seen as samples of a C^1 function f, then (with mild assumptions on $z_1, \ldots, z_N$) there is an underlying projector P with $\Gamma_P = \{z_1, \ldots, z_N\}$ and Im P(= S_P) is an N-dimensional subspace of C^1 spanned by the cardinal functions for the method. (These cardinal functions are closely analogous to, say, the fundamental Lagrange polynomials of Ex.1.1.)

This is, of course, already a solution to the problem of smooth interpolating surface construction. But it has the disadvantage that it is computationally a very expensive solution. Consequently, a second stage is applied (without ever resolving the first explicitly) in which the function Pf and its first derivatives are sampled on a *rectangular* grid of points in $\mathbb{R}^2$ which we denote by $\zeta_1, \ldots, \zeta_L$. From these samples a C^1 surface is generated using a finite element technique. This determines a second projector Q on C^1 with $\Gamma_Q = \zeta_1, \zeta_2, \ldots, \zeta_L$. The ultimate solution to the problem is the function QPf, and leads to analysis of the composition QP. However, unless $\Gamma_P \subset \Gamma_Q$ the function QPf will no longer interpolate on the set $\{z_1, \ldots, z_N\}$.

It is assumed here that either triangular or rectangular finite elements are used which have C^1 inter-element continuity and require the evaluation of derivatives with order no higher than one. Such elements certainly exist and examples suggest that local variations in the function Pf can give rise to large variations in second order derivatives which may then be transmitted to the support of all elements sharing such a node. The thesis of Susan Ritchie [12] reviews appropriate elements and contains numerical examples.

EXAMPLE 2.2 A variant of example 2.1 is to use the first step to generate *derivative* information at the given data points $z_1, z_2, \ldots, z_N$, or a subset of them. A triangulation of the plane is then determined with these data points as vertices, and using the given function values with derivative values from step one, a finite element method is used to generate the C^1 surface. There is again an underlying interpolating projector Q, of course.

In this approach, the problem of constructing a suitable triangulation has to be overcome. This has been considered by geographers, geophysicists and mathematicians independently (see, for example, Barnhill [1], Green and Sibson [7], Lawson [8], Rhynsburger [9]) and it seems that reliable software for achieving these triangulations will soon be generally available. We shall pay no further attention to this particular question.

Another variant is to define a projector Q on the original data, or a subset of it, in which case one can form Boolean sums $Q \oplus P = Q + P - QP$. Algorithms of this type are considered by Barnhill (p. 113 of [1]).

Return now to the general setting and let P, Q be any interpolating projectors from $L(X)$. What can be said about the interpolating sets and precision classes of QP and $Q + P - QP$? In general, one can make little more than the obvious statements:

$$(4) \qquad \Gamma_{QP} \supset \Gamma_Q \cap \Gamma_P , \qquad S_{QP} \supset S_P \cap S_Q ,$$

and for the Boolean sum the slightly less obvious statements (due to Barnhill and Gregory [2]),

$$(5) \qquad \Gamma_{Q \oplus P} \supset \Gamma_Q , \qquad S_{Q \oplus P} \supset S_P .$$

The latter follow from the factorization:

$$(6) \qquad I - (Q \oplus P) = I - P - Q + QP = (I-Q)(I-P) .$$

To make more informative statements one generally requires further properties of P and Q; preferably properties which are realised in some practical situations. For example, we shall shortly see an important example in which P and Q commute. Before stating the appropriate theorem, observe first that there is an identity dual to (6) obtained by replacing P, Q with $I - P$, $I - Q$:

$$(7) \qquad I - QP = (I-Q) \oplus (I-P) ,$$

for any projectors $P, Q \in L(X)$. This kind of duality is reflected strongly in Theorem 2 which it is safe to describe as "well-known".

THEOREM 2 *Let* $P, Q \in L(X)$ *and satisfy* $P^2 = P$, $Q^2 = Q$ *and* $PQ = QP$. *Then* QP *and* $Q \oplus P$ *are also projectors and*

$$\operatorname{Im} QP = (\operatorname{Im} Q) \cap (\operatorname{Im} P), \qquad \operatorname{Ker} QP = \operatorname{Ker} Q + \operatorname{Ker} P,$$
$$\operatorname{Im} (Q \oplus P) = \operatorname{Im} P + \operatorname{Im} Q, \qquad \operatorname{Ker} (P \oplus Q) = \operatorname{Ker} P \cap \operatorname{Ker} Q.$$

As further motivation for this theorem note that, since $X = \operatorname{Im} P \oplus \operatorname{Ker} P$ for *any* projector P, it can be argued that we can only understand P when $\operatorname{Im} P$ and $\operatorname{Ker} P$ are fully described.

The interpolation sets for projectors QP, $Q \oplus P$ of the theorem are also easily described (due to E.W. Cheney and W.J. Gordon, cf. [6]):

(8) $$\Gamma_{QP} = \Gamma_Q \cap \Gamma_P , \qquad \Gamma_{Q\oplus P} = \Gamma_Q \cup \Gamma_P .$$

Also, the formulae for $\operatorname{Im} QP$, $\operatorname{Im} Q \oplus P$ can be rewritten:

(9) $$S_{QP} = S_Q \cap S_P , \qquad S_{Q\oplus P} = S_Q + S_P .$$

These relations are, of course, to be compared with (4) and (5).

Gordon and Wixom [6] introduce the idea of *weakly commutative* projectors . This means that for any $f \in X$

$$PQf = QPf$$

at all points $z \in \Gamma_P \cup \Gamma_Q$. For such projectors they prove

(10) $$\Gamma_{QP} = \Gamma_Q \cap \Gamma_P , \qquad \Gamma_{Q\oplus P} \supset \Gamma_Q \cup \Gamma_P .$$

It is easily verified that projectors P, Q introduced in examples 2.1 and 2.2 are weakly commutative (but *not* commutative) so that these results will apply.

EXAMPLE 1.2 We form a product scheme on $\Omega = [a,b] \times [c,d]$ by combining processes like that introduced in Ex. 1.1. Let $C^{\mu,\nu}$ denote the class of functions f for which the derivatives $f^{(i,j)}$ exist and are uniformly continuous on $(a,b) \times (c,d)$ for $0 \le i \le \mu$, $0 \le j \le \nu$.

Let $\varphi_1, \ldots, \varphi_M$, $g_1, \ldots, g_M$, S be as defined in Ex. 1.1. In an analogous way, consider functionals $\psi_1, \ldots, \psi_N$ on $C^\nu[c,d]$ and construct a subspace T with dual T^* having dual bases $h_1, \ldots, h_N$ and $\psi_1, \ldots, \psi_N$.

The functional φ_i $(i = 1, \ldots, M)$ is used to define an operator Φ_i on $C^{\mu,\nu}$ by having φ_i act only on the x variable of any function $g(x,y) \in C^{\mu,\nu}$. A similar idea is used to determine an operator Ψ_j from functional ψ_j. Thus, for $i = 1, \ldots, M$,

$$\varphi_i \text{ determines } \Phi_i: \ C^{\mu,\nu}(\Omega) \to C^{\nu}[c,d]$$

and, for $j = 1, \ldots, N$,

$$\psi_j \text{ determines } \Psi_j: \ C^{\mu,\nu}(\Omega) \to C^{\mu}[a,b].$$

(Gordon [4] describes the Φ_i, Ψ_j as *parametric functionals.*) For example, if $g \in C^{\mu,\nu}(R)$ and $\varphi_i(f) = f(x_0)$ defines φ_i, then $\Phi_i g(x,y) = g(x_0,y) \in C^{\nu}[c,d]$.

We can then consider the formation of composite functionals $\varphi_i\Psi_j$ and $\psi_j\Phi_i$ acting on $C^{\mu,\nu}$. Let us make the assumption

$$(11) \qquad \varphi_i\Psi_j = \psi_j\Phi_i \ , \qquad \begin{cases} i = 1, \ldots, M, \\ j = 1, \ldots, N, \end{cases}$$

and write $\Theta_{ij} = \varphi_i\Psi_j$. Define operators from $C^{\mu,\nu}$ to itself by

$$(12) \qquad P_x = \sum_{i=1}^{M} g_i(x)\Phi_i \ , \qquad P_y = \sum_{j=1}^{N} h_j(y)\Psi_j \ ,$$

and let

$$U = \text{span}\ \{g_i(x)h_j(y)\} \ , \qquad \begin{cases} i = 1, \ldots, M \ , \\ j = 1, \ldots, N \ . \end{cases}$$

The following theorem is then, in some respects, an application of the general Theorem 2.

THEOREM 3 (a) *P_x, P_y are commuting projectors.*

(b) *If $f \in C^{\mu,\nu}(R)$ there is a unique $u \in U$ such that $\Theta_{rs}(u) = \Theta_{rs}(f)$, $1 \le r \le M$, $1 \le s \le N$ and, furthermore, $u = P_x P_y f$.*

(c) *If $f \in C^{\mu,\nu}(R)$ there is a unique $v \in$ Im P_x + Im P_y such that*

$$(13) \qquad (\Phi_r v)(y) = (\Phi_r f)(y), \qquad 1 \le r \le M \ ,$$

and

$$(14) \qquad (\Psi_s v)(x) = (\Psi_s f)(x), \qquad 1 \le s \le N \ ,$$

and, furthermore, $v = (P_x \oplus P_y)f$.

This theorem describes, in a quite general setting, the construction of tensor product interpolants (part (b)) and blended interpolants (part (c)). A comprehensive survey of this kind of construction is given by Gordon [4]. **See also C.A. Hall's first paper in this volume.**

It is important to note that P_x, P_y are projectors whose ranges are of infinite dimension. For example, speaking heuristically, P_x has the effect of restraining the x-behaviour of $P_x f$ to a space generated by the g_i, but the y-behaviour is still merely C^{ν}.

It should also be emphasized that there are MN discrete interpolation conditions satisfied by the tensor product interpolant u of part (b). In contrast, the conditions (13) and (14) for the blended interpolant v obtain on whole line segments. This has given rise to the description *transfinite* interpolation, [5].

To illustrate, suppose (as in Ex. 1.2) that the spaces $S \subset C^{\mu}[a,b]$ and $T \subset C^{\nu}[c,d]$ consist of natural cubic splines so that the bases $g_1(x), \ldots, g_M(x)$ and $h_1(y), \ldots, h_N(y)$ consist of the *cardinal* natural cubic splines. The functionals φ_i and ψ_j are then point functionals, $\varphi_i(f) = f(x_i)$, $i = 1, \ldots, M$, and so on.

Then $\Theta_{rs}g(x,y) = g(x_r,y_s)$ and the tensor product interpolant to g agrees with g at the vertices of the lattice generated by the two partitions $a = x_1 < x_2 < \ldots < x_M = b$ and $c = y_1 < y_2 < \ldots < y_N = d$. Indeed, P_xP_yg is what is now widely known as the *bicubic spline* interpolating to g on

(15) $$\Gamma_{P_xP_y} = \{(x_r,y_s): \; 1 \le r \le M, \; 1 \le s \le N\} .$$

In contrast, $(P_x \oplus P_y)g$ is the so-called *spline blended* interpolant and interpolates to g on the complete line segments of the lattice, i.e.

(16) $$\Gamma_{P_x \oplus P_y} = \{(x_r,y): \; 1 \le r \le M\} \cup \{(x,y_s): \; 1 \le s \le N\} .$$

Note, that as suggested by (8), the sets on the right are Γ_{P_x} and Γ_{P_y}, respectively. In (15) we have, of course, $\Gamma_{P_x} \cap \Gamma_{P_y}$.

EXAMPLE 2.3 Return now to the problem described as example 2.1. It is very easy to see that in general $PQ \neq QP$ so that Theorem 2 does not apply. However, as noted above, P and Q are weakly commutative so that (10) applies - although the first statement is relatively obvious and the second does not generally apply, since the data is not given to compute Qf. For the precision class, it seems that there is little more to be said than the second of equations (4).

However, a new feature appears if $\Gamma_P \subset \Gamma_Q$. In this case it is easy to verify that QP is a projector (even though $PQ \neq QP$). This property is also enjoyed by the first problem described in example 2.2, and for essentially the same reason.

Return once more to the abstract formalism and consider projectors P and Q with the property that QP is also a projector. The next theorem shows that this property admits a complete description of Im QP and Ker QP which generalizes the corresponding part of Theorem 2, and seems to be new.

THEOREM 4 *Let* $P, Q \in L(X)$ *and satisfy* $P^2 = P$, $Q^2 = Q$ *and* $(QP)^2 = QP$. *Then*

$$\text{(17)} \quad \text{Im}\,(QP) = \text{Im}\,Q \cap [\text{Im}\,P \oplus (\text{Ker}\,Q \cap \text{Ker}\,P)] \;,$$

$$\text{(18)} \quad \text{Ker}\,(QP) = \text{Ker}\,P \oplus (\text{Ker}\,Q \cap \text{Im}\,P) \;.$$

The proof of this theorem is by straight-forward methods of linear algebra and will not be given. Observe that QP a projection implies, using (7), that $(I-Q) \oplus (I-P)$ is a projection. However, PQ, $(I-Q)(I-P)$ and $Q \oplus P$ are not necessarily projections.

As an example of the application of Theorem 4, return to the two-stage surface fitting procedure.

EXAMPLE 2.4 Continuing from the discussion of Ex. 2.3 consider procedures (of the kind described in either 2.1 or 2.2) in which $\Gamma_P \subset \Gamma_Q$. It has been noted that QP is a projector so Theorem 4 applies. Since Ker Q (Ker P) can be described as the class of all C^1 functions which vanish at the nodes of the Q-projector (of the P-projector), it is easily seen that Ker $Q \subset$ Ker P. Consequently, Ker $Q \cap$ Im $P = \phi$ and (17), (18) reduce to

$$\text{Im}\,QP = \text{Im}\,Q \cap (\text{Im}\,P \oplus \text{Ker}\,Q)$$

$$\text{Ker}\,QP = \text{Ker}\,P \;.$$

The first of these equations allows us to describe a basis for Im QP as follows. Let $g_1, \ldots, g_N$ be the cardinal functions forming a basis for Im P. Then $Qg_1, \ldots, Qg_N$ are in Im Q and, because they retain the cardinal properties at $z_1, \ldots, z_N$, are linearly independent. Using the description of Ker Q above, one sees also, that since $Qg_1, \ldots, Qg_N$ retain the Q-nodal values of $g_1, \ldots, g_N$ they are in Im $P \oplus$ Ker Q. Consequently, they are in Im QP. Finally the dimension of Im QP cannot exceed that of Im P and so $Qg_1, \ldots, Qg_N$ form a basis for Im QP.

EXAMPLE 1.3 We conclude this paper by pursuing the question of the numerical implementation of blending interpolation, as described in Ex. 1.2, and showing that this also leads to problems in which Theorem 4 applies.

It has been noted that the projectors P_x, P_y constructed in Ex. 1.2 have infinite-dimensional range. Consequently, in order to represent a function $P_x f$ numerically one must, in general, consider an approximate representation of the y-dependence of the function

$$P_x f(x,y) = \Sigma\, g_i(x)\; \Phi_i f(.,y) \;.$$

This is usually accomplished by considering the image of $P_x f$ under a *second* projector, say $\overline{P}_y$, which acts on the y-variable of bivariate functions in $C^{\mu,\nu}$. Similarly, a projection $\overline{P}_x$ is introduced and $\overline{P}_x P_y f$ is used as an approximation for $P_y f$.

Assume that these new projectors have the same structure as P_x, P_y themselves. Thus, by comparison with equations (12), and omitting some obvious details in the definitions we write

$$\overline{P}_x = \sum_{k=1}^{\overline{M}} \overline{g}_k(x)\overline{\Phi}_k , \qquad \overline{P}_y = \sum_{l=1}^{\overline{N}} \overline{h}_l(y)\overline{\Psi}_l .$$

Since $\overline{P}_x$, $\overline{P}_y$ should be at least as precise as P_x, P_y it is reasonable to assume $\overline{M} \geq M$, $\overline{N} \geq N$.

One is then led to consider the behaviour of the transformation

$$(19) \qquad P = \overline{P}_y P_x + \overline{P}_x P_y - P_x P_y ,$$

with finite-dimensional range, as an approximation for the projector $P_x \oplus P_y$. These ideas are discussed by Gordon [4], and further developed by Cavendish, Gordon, and Hall, [3] and [5]. They reappear in Theorem 2 of Hall's first paper in this volume.

Assuming the functionals $\varphi_1, \ldots, \varphi_M$ to be defined by nodes of interpolation, it is not unnatural to define $\overline{\varphi}_1, \ldots, \overline{\varphi}_{\overline{M}}$ as a refinement of the first set of nodes. Thus, $\{\varphi_i\} \subset \{\overline{\varphi}_k\}$. The associated function space may also be an extension of that for P_x, i.e., span $\{g_i\} \subset$ span $\{\overline{g}_k\}$. Similar inclusions arise in the same natural way for P_y and $\overline{P}_y$. Thus, $\{\psi_j\} \subset \{\overline{\Psi}_l\}$ and/or span $\{h_j\} \subset$ span $\{\overline{h}_l\}$. Consider the inclusions of nodes and subspaces separately:

$$(20) \qquad \{\varphi_i\} \subset \{\overline{\varphi}_k\} \quad \text{and} \quad \{\psi_j\} \subset \{\overline{\psi}_l\} ,$$

$$(21) \qquad \text{span}\ \{g_i\} \subset \text{span}\ \{\overline{g}_k\} \quad \text{and} \quad \text{span}\ \{h_j\} \subset \text{span}\ \{\overline{h}_l\} .$$

It is also natural and generally no restriction to extend the commutativity conditions formulated in Ex. 1.2:

$$(22) \qquad \varphi_i\overline{\Psi}_l = \overline{\psi}_l\Phi_i \quad \text{and} \quad \overline{\varphi}_k\Psi_j = \psi_j\overline{\Phi}_k$$

for $1 \leq i \leq M$, $1 \leq j \leq N$, $1 \leq k \leq \overline{M}$, $1 \leq l \leq \overline{N}$.

The following theorem can then be proved which clarifies some work of Cavendish, Gordon, and Hall [3]:

THEOREM 5 (a) *Hypotheses* (20) *and* (22) *imply that, for the operator* P *of* (19),

$$P = \left(P_x \oplus P_y\right)\left(\overline{P}_x\overline{P}_y\right) \quad \text{and} \quad P^2 = P .$$

(b) *Hypotheses* (21) *and* (22) *imply that*

$$P = (\overline{P}_x\overline{P}_y)(P_x \oplus P_y) \quad and \quad P^2 = P\ .$$

Observe that, if (20), (21) and (22) all hold, then it follows that $\overline{P}_x\overline{P}_y$ and $P_x \oplus P_y$ commute and Im P, Ker P are given by Theorem 2. Under the conditions of part (a) or of part (b), however, Theorem 4 must be used to obtain the same information.

It is easily seen that, in part (a) the relations

$$P_x\overline{P}_x = P_x \qquad \text{and} \qquad P_y\overline{P}_y = P_y$$

hold and in part (b), Im $P_x \subset$ Im $\overline{P}_x$, Im $P_y \subset$ Im $\overline{P}_y$, so that

$$\overline{P}_xP_x = P_x \qquad \text{and} \qquad \overline{P}_yP_y = P_y\ .$$

Cavendish, Gordon and Hall describe the latter conditions as *absorptive*.

The following example demonstrates some of the major features of Theorems 4 and 5 and the arguments are similar to those used by Watkins and Lancaster [14] in the construction of finite elements. Let $R = [-1,1] \times [-1,1]$, $P_xu = u(0,y)$, $P_y(u) = u(x,0)$ so that

$$(P_x \oplus P_y)u = u(x,0) + u(0,y) - u(0,0)\ .$$

This is a blended projector on R whose interpolation set is $\{(x,y): x = 0 \text{ or } y = 0\}$. The underlying set of functionals of one variable is the singleton $\{\varphi_1\}$ where $\varphi_1(f) = f(0)$.

Then define $\overline{\varphi}_1(f) = f(-1)$, $\overline{\varphi}_2(1) = f(1)$ and corresponding projectors

$$\overline{P}_xu = \tfrac{1}{2}(1-x)u(-1,y) + \tfrac{1}{2}(1+x)u(1,y)\ ,$$

$$\overline{P}_yu = \tfrac{1}{2}(1-y)u(x,-1) + \tfrac{1}{2}(1+y)u(x,1)\ .$$

With these definitions

$$P^0 = \text{span}\ \{g_1\} \subset \text{span}\ \{\overline{g}_1,\overline{g}_2\} = P^1$$

but $\{\varphi_1\} \not\subset \{\overline{\varphi}_1\ \overline{\varphi}_2\}$. Thus, case (b) of the theorem applies. It is found that

$$\begin{aligned}(\overline{P}_x\overline{P}_y)(P_x \oplus P_y)u &= \tfrac{1}{2}(1+x)u(1,0) + \tfrac{1}{2}(1+y)u(0,1)\\ &\quad + \tfrac{1}{2}(1-x)u(-1,0) + \tfrac{1}{2}(1-y)u(0,-1) - u(0,0)\ ,\end{aligned}$$

$$\begin{aligned}(P_x \oplus P_y)(\overline{P}_x\overline{P}_y)u &= \tfrac{1}{4}\,[(1+x+y)u(1,1) + (1-x+y)u(-1,1)\\ &\quad + (1-x-y)u(-1,-1) + (1+x-y)u(1,-1)]\ ,\end{aligned}$$

demonstrating the lack of commutativity of any kind.

It is easy to see that Im $P_x \oplus P_y$ and Im $\overline{P}_x\overline{P}_y$ both contain the functions 1, x, y so

$$\text{Im}\left(\overline{P}_x\overline{P}_y\right)\left(P_x \oplus P_y\right) \supset \text{span}\,\{1,x,y\}\ .$$

Now use the characterization (17): A function v, with

$$v \in \left(\text{Im}\, P_x \oplus P_y\right) \oplus \{\left(\text{Ker}\, P_x \oplus P_y\right) \cap \text{Ker}\, \overline{P}_x\overline{P}_y\}$$

has the form

$$v(x,y) = f(0,y) + g(x,0) + h(x,y)$$

for C^1 functions f, g and h with $h(0,y) = h(x,0) = 0$ in R and h zero at the vertices of R. But if v is also in Im $\overline{P}_x\overline{P}_y$ it must also satisfy $\overline{P}_x\overline{P}_y v = v$. This is found to imply that $v = \overline{P}_x\overline{P}_y v \in \text{span}\,\{1,x,y\} = P^1$. Thus Im $(\overline{P}_x\overline{P}_y)(P_x \oplus P_y)$ is just P^1.

ACKNOWLEDGEMENT

Some of the ideas of this paper were presented in the Seminar of Prof. E. Bohl at the University of Münster in June of 1978. The author is grateful to Prof. Bohl, the University of Münster, and the Deutsche Forschungsgemeinschaft for making this possible.

REFERENCES

[1] R.E. Barnhill. *Representation and approximation of surfaces.* Mathematical Software III, Academic Press, 1977, pp. 69-120.

[2] R.E. Barnhill and J.A. Gregory. *Polynomial interpolation to boundary data on triangles,* Math. Comp. 29 (1975), pp. 726-735.

[3] J.C. Cavendish, W.J. Gordon, and C.A. Hall. *Ritz-Galerkin approximation in blending function spaces,* Numer. Math. 26 (1976), pp. 155-178.

[4] W.J. Gordon. *Distributive lattices and the approximation of multivariate functions.* Approximations with Special Emphasis on Spline Functions, Academic Press, 1969, pp. 223-277.

[5] W.J. Gordon and C.A. Hall. *Transfinite element methods: Blending function interpolation over arbitrary curved element domains,* Numer. Math. 21 (1973), pp. 109-129.

[6] W.J. Gordon and J.A. Wixom. *Pseudo-harmonic interpolation on convex domains,* SIAM J. Numer. Anal. 11 (1974), pp. 909-933.

[7] P.J. Green and R. Sibson. *Computing Dirichlet tesselations in the plane,* Computer J., 21 (1978), pp. 168-173.

[8] P. Lancaster and K. Salkauskas. *A Survey of Curve and Surface Fitting.* Published by the authors, Calgary, 1977.

[9] C.L. Lawson. *Software for C^1 surface interpolation.* Mathematical Software III, Academic Press, 1977, pp.69-120.

[10] M.J.D. Powell. *Numerical Methods for fitting functions of two variables,* The State of the Art in Numerical Analysis, ed. D. Jacobs, Academic Press, 1977, pp. 563-604.

[11] D. Rhynsburger. *Analytic delineation of Thiessen polygons,* Geograph. Anal. 5 (1973), pp. 133-144.

[12] S. Ritchie, *Surface Representation by Finite Elements.* M.Sc. Thesis, Dept. of Math. and Stat., Univ. of Calgary, 1978.

[13] L.L. Schumaker. *Fitting surfaces to scattered data.* Approximation Theory II, Academic Press, 1976, pp. 203-268.

[14] D.S. Watkins and P. Lancaster. *Some families of finite elements,* J. Inst. Math. Applics. 19 (1977), pp. 385-397.

MOVING WEIGHTED LEAST-SQUARES METHODS

Peter Lancaster
The University of Calgary

ABSTRACT

Moving weighted least squares are defined and analyzed with a view to clarifying their algebraic structure and imposing sufficient conditions to guarantee interpolating and smoothness properties of the resulting surface.

1. PRELIMINARIES

Let $z = (x,y)$ denote a typical point of the Euclidean plane. Given data values f_i at distinct points $z_i = (x_i,y_i)$, $i = 1,2,\ldots,N$ a surface with height $f(z)$ at z is to be generated with the following properties: (a) $f(z_i) = f_i$, $i = 1,\ldots,N$ and, (b) the function f is at least C^1. The domain of f may be the whole plane although, in applications, it is generally a finite simply-connected domain containing the convex hull of the set $\{z_i\}_{i=1}^N$.

There are many practical situations in which such surfaces are useful - in geography, geophysics, engineering, and so on, but it is important to note at the start that for any given $\{z_i\}_{i=1}^N$ there are infinitely many surfaces having the required properties of interpolation and smoothness. Furthermore the criteria for a "good" solution are generally not well understood by those posing the problem. Although some progress is being made toward objective criteria for defining desirable surfaces more clearly, such criteria will not be considered here and we present just one class of methods to generate surfaces with properties (a) and (b), together with examples and analysis.

The processes to be discussed have been described as "moving least squares" methods and there is some literature to be found. For example, there is considerable numerical experience summarized in the work of McLain [4] and [5], and there are some foundations laid by Barnhill and his group [1], Gordon and Wixom [2], Schumaker [7], and Ritchie [6]. Much of what is to be said here will also apply to univariate

Badri N. Sahney (ed.), Polynomial and Spline Approximation, 103–120.

or multivariate data, but we shall stay with just two space variables for clarity.

The least squares technique to be developed will employ a linearly independent set of real-valued functions $\{b_i(z)\}_{i=1}^n$, $n \le N$, and it is convenient at this stage to formulate the classical least squares fitting procedure. If $S = \text{span } \{b_i(z)\}_{i=1}^n$, a function in S is to be determined so that the mean-square error at the data points is minimized. Thus, the real coefficients $\alpha_1,\ldots,\alpha_n$ are to be determined so as to minimize

$$(1) \qquad E(\alpha_1,\ldots,\alpha_n) = \sum_{j=1}^{N}\left\{\left(\sum_{i=1}^{n} \alpha_i b_i(z_j)\right) - f_j\right\}^2$$

The necessary conditions for E to have a stationary value yield the normal equations to be solved for $\alpha_1,\ldots,\alpha_n$. Writing these as a column vector $\underset{\sim}{\alpha}$ the normal equations have the form

$$(2) \qquad \hat{A}\underset{\sim}{\alpha} = \hat{\underset{\sim}{\varphi}}$$

where, for $r,s = 1,2,\ldots,n$,

$$(3) \qquad \hat{a}_{rs} = \sum_{j=1}^{N} b_r(z_j)b_s(z_j), \quad \hat{\varphi}_r = \sum_{j=1}^{N} b_r(z_j)f_j.$$

Note that $\hat{A}$ is real and symmetric but *could* be singular if, in in particular, $\{z_j\}_{j=1}^N$ contains no subset of n points unisolvent with respect to $\{b_i(z)\}_{i=1}^n$. When a unique solution of $\hat{A}\underset{\sim}{\alpha} = \hat{\underset{\sim}{\varphi}}$ exists this determines the "best least squares" fit to the data from S and has the form

$$\sum_{i=1}^{n} \alpha_i b_i(z).$$

In general, this function will *not* interpolate the data, and it will have the smoothness of the basis functions $\{b_i(z)\}_{i=1}^n$.

If some pieces of data are thought to be more reliable or more important than others, a surface can be generated which will give smaller errors at selected abscissas. This is done by introducing weights $w_j > 0$, $j = 1,2,\ldots,N$, associated with the abscissas z_j and replacing (1) by:

$$(4) \qquad E(\alpha_1,\ldots,\alpha_n) = \sum_{j=1}^{N} w_j\left\{\left(\sum_{i=1}^{n} \alpha_i b_i(z_j)\right) - f_j\right\}^2$$

In this case the normal equations (2) have coefficients:

$$(5) \qquad \hat{a}_{rs} = \sum_{j=1}^{N} w_j b_r(z_j)b_s(z_j), \quad \hat{\varphi}_r = \sum_{j=1}^{N} w_j b_r(z_j)f_j .$$

2. FORMULATION OF THE DISTANCE-WEIGHTED LEAST SQUARES METHOD

The technique of the title of this paper is motivated by the following consideration: In determining the surface height f at some point $\hat{z} \neq z_i$, $i = 1,2,\ldots,N$, the data values f_i at points near to $\hat{z}$ should be given greater weight than more distant points. Thus, to determine f at $\hat{z}$, first find (implicitly) a function $F(\hat{z};\cdot)$ in S by the weighted least squares method:

$$(6) \qquad F(\hat{z};z) = \sum_{i=1}^{n} \alpha_i(\hat{z})b_i(z) = \underset{\sim}{\alpha}(\hat{z})^T \underset{\sim}{b}(z),$$

where a positive weight function $w(\hat{z};z)$ is used to assign the weights $w_j = w(\hat{z};z_j)$, $j = 1,2,\ldots,N$, required in the formulation of (4) and (5). A typical and popular choice of weight function (particularly convenient for analysis) is

$$(7) \qquad w(\hat{z};z) = \{(x-\hat{x})^2 + (y - \hat{y}^2)\}^{-1},$$

or, $w(\hat{z};z) = d^{-2}(z,\hat{z})$ where $d(z,\hat{z})$ is the euclidean distance from z to $\hat{z}$.

The value of the surface height at $\hat{z}$ is then defined by $f(\hat{z}) = F(\hat{z};\hat{z})$ and F is defined by (6), the coefficients $\alpha_i(\hat{z})$ are in turn determined by the solution $\underset{\sim}{\alpha}$ of

$$(8) \qquad \tilde{A}(\hat{z})\underset{\sim}{\alpha} = \underset{\sim}{\tilde{\varphi}}(\hat{z}),$$

where for $r,s = 1,2,\ldots,n$ (cf.equation (5)):

$$(9) \qquad \tilde{a}_{rs}(\hat{z}) = \sum_{j=1}^{N} w(\hat{z};z_j)b_r(z_j)b_s(z_j), \quad \tilde{\phi}_r(\hat{z}) = \sum_{j=1}^{N} w(\hat{z};z_j)b_r(z_j)f_j.$$

Let B be the $N \times n$ matrix with elements $b_r(z_j)$ on row j and column r (i.e. with rows $\underset{\sim}{b}^T(z_j)$) and

$$W(\hat{z}) = \operatorname{diag}\{w(\hat{z},z_1),\ldots,w(\hat{z},z_N)\},$$

then the coefficient matrix $\tilde{A}(\hat{z})$ of (8) has the factorization

$$(10) \qquad \tilde{A}(\hat{z}) = B^T W(\hat{z})B.$$

From this factorization it is apparent that $\tilde{A}(\hat{z})$ *is positive definite if and only if there is a subset of n points in* $\{z_i\}_{i=1}^N$ *which is unisolvent with respect to* $\{b_i(z)\}_{i=1}^n$. When this is the case, we say the problem is *well posed* and the procedure described above will determine uniquely the height $f(\hat{z})$ for any $\hat{z} \notin \{z_i\}_{i=1}^N$

The procedure breaks down at the data points because $w(\hat{z};z_i)$ is not well-defined by (7) when $\hat{z} = z_i$. Stated another way, the coefficients of $\tilde{A}$, $\tilde{\varphi}$ in (8), seen as functions of $\hat{z}$, have singularities at the data points. The nature of this difficulty is simplified if equations (8) are scaled by multiplying left and right by the nonnegative function of $\hat{z}$:

$$\prod_{i=1}^{N} d^2(z_i,\hat{z}) \Big/ \sum_{i=1}^{N} \left\{ \prod_{\substack{k=1\\k\neq i}}^{N} d^2(z_k,\hat{z}) \right\}.$$

Since this function vanishes if and only if $\hat{z} \in \{z_i\}_{i=1}^N$, the coefficient matrix retains the positive definite property. Replace now $\hat{z}$ by z and denote the scaled matrix and vector by $A(z)$, $\underset{\sim}{\varphi}(z)$ so that the defining equations for the process are as follows:

$$A(z)\underset{\sim}{a} = \underset{\sim}{\varphi}(z), \tag{11}$$

$$a_{rs}(z) = \sum_{j=1}^{N} v_j(z)b_r(z_j)b_s(z_j), \quad \varphi_r(z) = \sum_{j=1}^{N} v_j(z)b_r(z_j)f_j, \tag{12}$$

for $r,s = 1,2,\ldots,n$, and, with some manipulation,

$$v_j(z) = \{1 + d^2(z,z_j)\sum_{\substack{i=1\\i\neq j}}^{N} d^{-2}(z,z_i)\}^{-1}. \tag{13}$$

With this formulation, the functions v_j can be defined at the points of $\{z_i\}_{i=1}^N$ by continuity, and so also the coefficients of equation (11). The normalized weight functions $v_j(z)$ play an important role in the sequel so we list here some of their more obvious properties:

$$\left\{\begin{array}{ll} v_i(z_j) = \delta_{ij}\,, & i,j = 1,2,\ldots,N \\ 0 \le v_j(z) \le 1 & \forall z \text{ and } v_j(z) = 0 \text{ iff } z = z_i,\ i \neq j\,. \\ \sum_{j=1}^{N} v_j(z) = 1 & \forall z \\ v_j(z) \to 1/N & \text{as } \|z\| \to \infty \end{array}\right. \tag{14}$$

Writing $\underset{\sim}{b}(z)$ for the n-vector function with elements $b_1(z),\ldots,b_n(z)$ (as in (6)), it follows from the first of properties (14) that (see also equation (10)):

$$A(z_j) = \underset{\sim}{b}(z_j)\underset{\sim}{b}^T(z_j) \qquad \underset{\sim}{\varphi}(z_j) = \underset{\sim}{b}(z_j)f_j\,. \tag{15}$$

Thus, after normalization of (8) and continuation of A to the data points, singularity of A at these points has been traded for severe *rank deficiency* of A at the same points. However, it is clear from (15) that solutions to the algebraic systems

do exist at the data points. The question which is to occupy us is roughly as follows:
Given the surface $f(z)$ defined by

$$A(z)\,\underset{\sim}{\alpha}(z) = \underset{\sim}{\varrho}(z), \qquad f(z) = \underset{\sim}{\alpha}^T(z)\,\underset{\sim}{b}(z) \tag{16}$$

for all $z \notin \{z_i\}_{i=1}^N$ and by continuity at the set $\{z_i\}_{i=1}^N$, does it meet the criteria (a) and (b) of our opening paragraph?

3. DISCUSSION OF THE METHOD

A relatively well-known method is obtained in the special case $n = 1$, $b_1(z) \equiv 1$. The normal equations of (16) reduce to a single equation and the solution is

$$f(z) = \alpha_1(z) = \left\{\sum_{j=1}^{N} v_j(z) f_j\right\} / \sum_{j=1}^{N} v_j(z)\ . \tag{17}$$

This is precisely one of the methods proposed by Shepard [8], and now known as Shepard's method ([1] and [2]). Now the functions $v_j(z)$ are clearly rational functions in x and y (see eq. (13)) and have no singularities. They are consequently C^∞ and, since

$$\sum_{j=1}^{N} v_j(z) \neq 0$$

for any z, the function f of (17) is also C^∞. Indeed, one sees immediately from (17) and (14) that $\{v_j(z)\}_{j=1}^N$ are precisely *cardinal functions* for Shepard's method, i.e. $v_j(z)$ is the Shepard interpolant defined by data $f_i = \delta_{ij}$, $i = 1,2,\ldots,N$.

Defining $M_1 = \text{span}\{v_j(z)\}_{j=1}^N$ and interpreting the data set $\{f_j\}_{j=1}^N$ as values of some C^0 function, say, then Shepard's method determines a projection, S, of C^0 onto the N-dimensional subspace M_1. Remarks to this effect can also be found in the papers mentioned above. To what extent do they generalize when $n > 1$?

First, it is not difficult to see that the components of $\underset{\sim}{\alpha}(z)$, i.e. $\underset{\sim}{\alpha}(x,y)$, are rational functions of x and y, but as we shall see they may have singularities (perhaps "removable" singularities) at the points of $\{z_j\}_{j=1}^N$, which do not allow us to deduce the existence of continuous derivatives at the data points. Even so the solution f is certainly C^∞ on any domain not containing a data point.

The notion of a cardinal basis is still useful. For each j let cardinal functions $c_j(z)$ be defined by providing data sets $f_i = \delta_{ij}$, $i = 1,2,\ldots N$. Then it is apparent from equations (11) and (12) that, at least on $\mathbb{R}^2\backslash\{z_i\}_{i=1}^N$ we have

(18) $f(z) = \sum_{j=1}^{N} f_j c_j(z)$

Indeed, if $f(z_j)$ can be defined by continuity for each j, then the general process with n basis functions also determines a projector from C^0 onto an N-dimensional space, M_n say. In the general case, $M_n = \operatorname{span}\{c_j(z)\}_{j=1}^{N}$. (*)

In discussing the smoothness of interpolants in general, this discussion allows us to confine attention to the smoothness of the cardinal functions.

Note that at points not in the set $\{z_i\}_{i=1}^{N}$ we have

(19) $\frac{\partial f}{\partial x} = \frac{\partial \alpha^T}{\partial x} b + \alpha^T \frac{\partial b}{\partial x}$

and for the computation of α and $\frac{\partial \alpha}{\partial x}$, $\alpha = A^{-1} \varphi$ so that

$$\frac{\partial \alpha}{\partial x} = - A^{-1} \frac{\partial A}{\partial x} A^{-1} \varphi + A^{-1} \frac{\partial \varphi}{\partial x}$$

(20) $\qquad = A^{-1}\{- \frac{\partial A}{\partial x} \alpha + \frac{\partial \varphi}{\partial x}\}.$

This means that the evaluation of α, $\frac{\partial \alpha}{\partial x}$ and $\frac{\partial \alpha}{\partial y}$ (for example) requires the successive solution of three sets of equations $Ax = u_i$, with different right-hand side vectors u_1, u_2, u_3.

The algebraic structure of the problem is clarified by introducing the matrix B of equation (10) and matrix

$$V(z) = \operatorname{diag}\{v_1(z), \ldots, v_N(z)\} .$$

Then $A(z) = B^T V(z) B$, $\varphi(z) = B^T V f$, where $f^T = [f_1, \ldots, f_N]$, and the basic equation of (16) is

(21) $B^T V(z) B \alpha = B^T V(z) f$.

From this it is easy to see that

(22) $M_n \supset \operatorname{span}\{b_i(z)\}_{i=1}$.

For, with data specified by the rth basis function, $b_r(z)$, we have $f = B_{*r}$, the rth column of matrix B. Then it is apparent that (with a well-posed problem) (21) has the unique solution $\alpha = e_r$ (the rth unit vector in $\mathbb{R}^n$), so that in (16) $f(z) = e_r^T b(z) = b_r(z)$. This shows that for $r = 1, 2, \ldots, n$, $b_r(z) \subset M_n$, as required.

(*) The author is indebted to K. Salkauskas for this observation.

Another important observation concerns the interpolation property. We have noted that, at a data point, the coefficient matrix of (21) is singular. However, solutions exist and *every solution vector* $\underset{\sim}{\alpha}$ *of* (21) *at the point* $z = z_j$ *satisfies the condition* $\underset{\sim}{\alpha}^T \underset{\sim}{b}(z_j) = f_j$ *provided that*

$$(23) \qquad \underset{\sim}{b}(z_j) \neq \underset{\sim}{0}, \quad (j = 1,2,\ldots,N)\ .$$

To see this observe that $v_i(z_j) = \delta_{ij}$ so that equation (21) reduces to $\underset{\sim}{b}(z_j)(\underset{\sim}{b}^T(z_j)\underset{\sim}{\alpha}) = \underset{\sim}{b}(z_j)f_j$, as required.

In order to get a "feeling" for the method it is helpful to see some examples. The most important choice of basis functions $\{b_i(z)\}_{i=1}^n$ is certainly a basis of polynomial functions. For example with $n = 1$ and $b_1(z) = 1$ we have a basis for the space P^0 consisting of constant functions. This choice yields Shepard's method already mentioned.

With $n = 3$, $b_1(z) \equiv 1$, $b_2(z) = x$, $b_3(z) = y$ we have $\text{span}\{b_i(z)\}_{i=1}^3 = P^1 \subset M_3$, and so on.

Computational experiments with such polynomial basis functions (and $n = 1,3,4,6,8,10,15$) are reported by McLain [4]. Gordon and Wixom [2] make brief reference to the $n = 3$ case mentioned above. In fact, there is quite a large body of computational experience suggesting that the methods under discussion include practical and useful techniques.

As previous authors have pointed out there are many possible choices for the weight function. The inverse square distance function of (7) is used here mainly for convenience. The nature of the surface generated seems not to be very sensitive to the choice of w provided there is a singularity in $w(\hat{z};z)$ as $z \to \hat{z}$ and $w(\hat{z},z) \to 0$ sufficiently fast as $z \to \infty$.

The first two numerical examples were deliberate, and unsuccessful, attempts to produce a non-smooth surface. Hence the unusual choice of basis functions.

EX.1. <u>A univariate example</u>. Let $n = 2, b_1(x) = 1$, $b_2(x) = x^2$ and $n = 3$ with $x_1 = -1$, $x_2 = 0$, $x_3 = 1$.

It is found that $\alpha_1(x) = f_2$,

$$\alpha_2(x) = \frac{(x-1)^2}{2(x^2+1)} f_1 - f_2 + \frac{(x+1)^2}{2(x^2+1)} f_3\ .$$

Then the cardinal basis functions are:

$$c_1(x) = \frac{x^2(x-1)^2}{2(x^2+1)}, \quad c_2(x) = 1 - x^2, \quad c_3(x) = \frac{x^2(x+1)^2}{2(x^2+1)}\ .$$

These are sketched in Fig. 1.

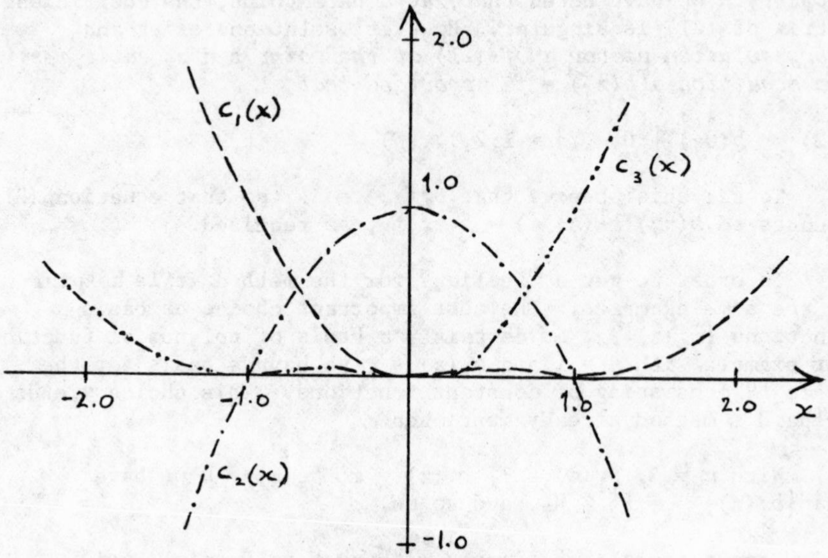

Fig. 1 Cardinal functions for a univariate example.

EX. 2. <u>A bivariate example</u>. Let $n = 2$, $b_1(z) = 1$, $b_2(z) = x$ and $N = 3$. $z_1 = (0,0)$, $z_2 = (1,0)$, $z_3 = (0,1)$.

Writing $d_i^2 = (x-x_i)^2 + (y-y_i)^2$, $i = 1,2,3$ it is found that

$$\alpha_1(x,y) = (d_3^2 f_1 + d_1^2 f_3)/(d_1^2+d_3^2),$$

$$\alpha_2(x,y) = \left[d_3^2(f_2-f_1) + d_1^2(f_2-f_3)\right]/(d_1^2+d_3^2).$$

Then the cardinal functions are:

$$c_1(x,y) = d_3^2(1-x)/(d_1^2+d_3^2),\ c_2(x,y) = x\ ,\ c_3(x,y) = d_1^2(1-x)/(d_1^2+d_3^2).$$

The first of these functions is indicated in Fig. 2.

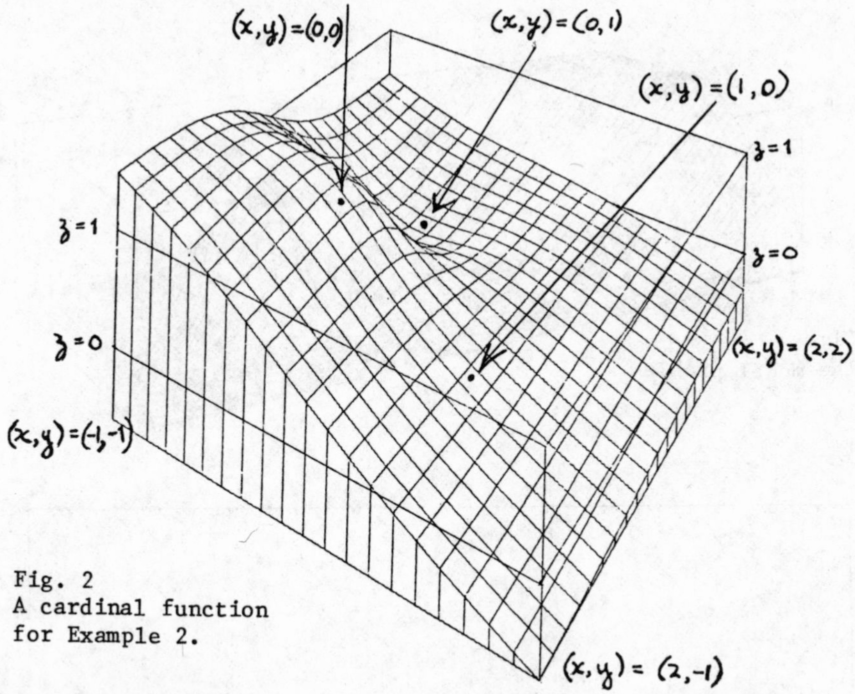

Fig. 2
A cardinal function
for Example 2.

EX.3 <u>A Model Problem</u>. Let $n = 6$ and $b_1(z), \ldots, b_6(z)$ be the usual basis for P^2. In this case a mathematically defined function made up of planes and a "mountain", illustrated in Fig. 3, is defined on a rectangular domain. Data points are generated by the random selection of 150 data points in the rectangle (Fig. 4) and the function sampled at these points to determine the data f_j, $j = 1, \ldots, 150$.

The resulting moving least squares surface is illustrated in Fig. 5.

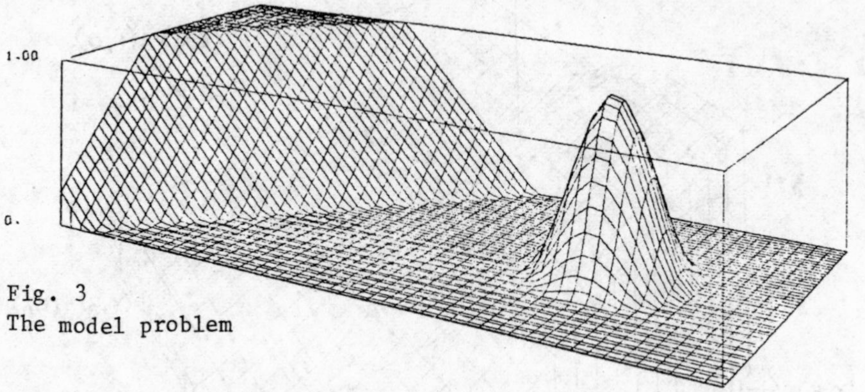

Fig. 3
The model problem

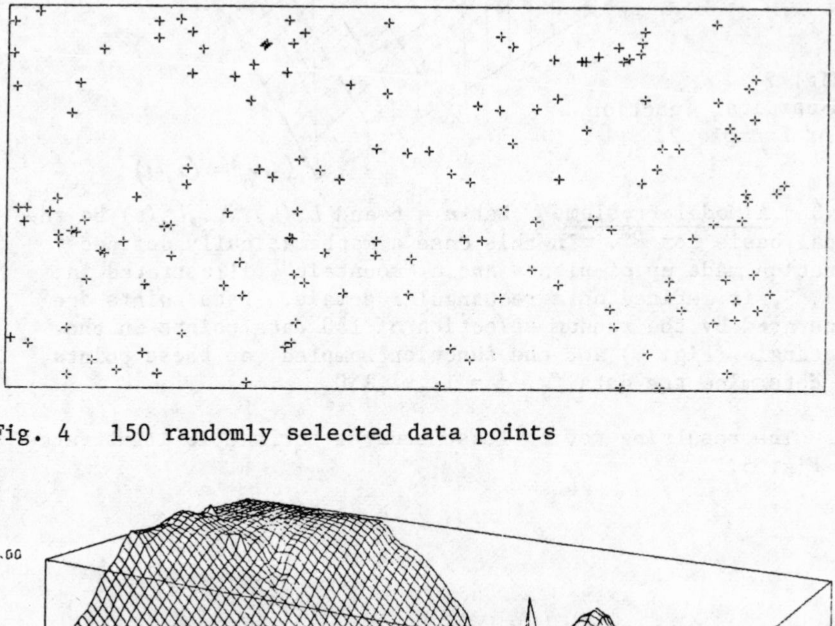

Fig. 4 150 randomly selected data points

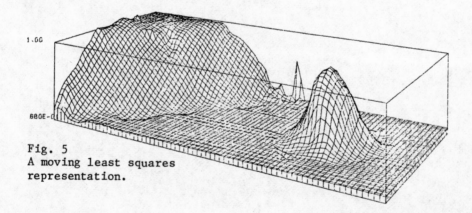

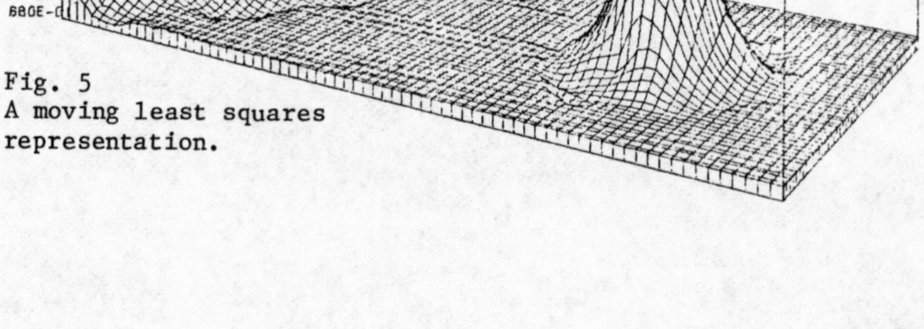

Fig. 5
A moving least squares
representation.

4. PRELIMINARY ANALYSIS

The remainder of this paper is devoted to a proof of the C^1 continuity of the methods defined under a sufficient condition to be enunciated shortly. As we have observed, it is only at the data points where we have anything to prove for the surface is necessarily C^∞ on any domain not containing data points. It should be pointed out, however, that the *necessity* of our condition is not clear and, to the time of writing, the author has not found an explicit example in which a surface defined by the methods of this paper fails to have at least C^1 smoothness on the whole of $\mathbb{R}^2$.

LEMMA 1 *The normalized weight functions of equations* (12) *and* (13) *have the following properties for* $k = 1,2,\dots,N$:

(i) $v_k(z) \in C^\infty$

(ii) $\left.\dfrac{\partial v_k}{\partial x}\right|_{z_j} = \left.\dfrac{\partial v_k}{\partial y}\right|_{z_j} = 0, \qquad j = 1,2,\dots,N$.

(iii) *If r denotes the euclidean distance from z to z_k, and $d_{j,k}$ the euclidean distance from z_j to z_k then as $z \to z_k$* (i.e. as $r \to 0$),

$$(24) \qquad v_k(z) = 1 - r^2\left\{\sum_{\substack{j=1\\ j\neq k}}^{N} d_{j,k}^{-2}\right\} + 0(r^3),$$

and for $j \neq k$,

$$(25) \qquad v_j(z) = r^2(d_{j,k}^{-2}) + 0(r^3).$$

Proof. Without loss of generality one may assume $z_k = 0$ and then write $v_k(z) = v(z)$. It has been remarked that this function is C^∞ on $\mathbb{R}^2$ so Taylor's theorem can be used to obtain the results. Define

$$\vartheta_j(z) = d^2(z,z_k)/d^2(z,z_j) = (x^2+y^2)/\{(x-x_j)^2+(y-y_j)^2\}$$

and from (13) we have

$$v_k(z) = v(z) = \left\{1 + \sum_{j=1}^{M} \vartheta_j(z)\right\}^{-1}$$

It is easily seen that at (0,0) we have $\vartheta_j = \dfrac{\partial\vartheta_j}{\partial x} = \dfrac{\partial\vartheta_j}{\partial y} = 0$ and

$$\frac{\partial^2\vartheta_j}{\partial x^2} = \frac{\partial^2\vartheta_j}{\partial y^2} = 2d^{-2}(0,z_j), \qquad \frac{\partial^2\vartheta_j}{\partial x\partial y} = 0$$

Differentiating the relation $\{1 + \sum\vartheta_j(z)\}v(z) = 1$ successively and evaluating derivatives at $x = y = 0$ it is then found that at the origin,

$$v = 1, \quad \frac{\partial v}{\partial x} = \frac{\partial v}{\partial y} = \frac{\partial^2 v}{\partial x \partial y} = 0, \qquad \frac{\partial^2 v}{\partial x^2} = \frac{\partial^2 v}{\partial y^2} = -2 \sum_{j=1}^{N} d^{-2}(0, z_j),$$

and that there are non-zero third order derivatives at the origin. Equation (24) then follows from Taylor's theorem.

When $j \neq k$ equation (13) is manipulated to give

$$v_j(z) = \frac{r^2}{d^2(z,z_j) + r^2 \sum_{i \neq k} \frac{d^2(z,z_j)}{d^2(z,z_i)}}$$

and (25) follows from this.//

It follows immediately from lemma 1 that $v_k(z)$ has a maximum at z_k and a minimum at z_j, for $j = 1,2,\ldots,N$, $j \neq k$. Note that - as is obvious - the functions $v_k(z)$, $v_j(z)$ cannot depend only on r. The terms destroying the radial symmetry are simply confined to the $0(r^3)$ terms.

Now apply the lemma to examine the behaviour of $A(z)$, $\underset{\sim}{\varphi}(z)$ of equation (16) in a neighbourhood of z_k. Define $N \times N$ diagonal matrices V_0, with a one in the k,k position and zeros elsewhere, and

$$(26) \qquad V_2 = \operatorname{diag}\{d_{1,k}^{-2}, \ldots, d_{k-1,k}^{-2}, -\sum_{j \neq k} d_{j,k}^{-2}, d_{k+1,k}^{-2}, \ldots, d_{N,k}^{-2}\}$$

Then in equation (21) we have $V(z) = V_0 + r^2 V_2 + 0(r^3)$ and in (16), as $z \to z_k$ (i.e. as $r \to 0$),

$$(27) \qquad A(z) = A_0 + r^2 A_2 + 0(r^3) \ ,$$

$$(28) \qquad \underset{\sim}{\varphi}(z) = \underset{\sim}{\varphi}_0 + r^2 \underset{\sim}{\varphi}_2 + 0(r^3) \ ,$$

where

$$(29) \qquad A_0 = B^T V_0 B = \underset{\sim}{b}(z_k)\underset{\sim}{b}^T(z_k), \qquad A_2 = B^T V_2 B,$$

$$(30) \qquad \underset{\sim}{\varphi}_0 = B^T V_0 \underset{\sim}{f} = f_k \underset{\sim}{b}(z_k), \qquad \underset{\sim}{\varphi}_2 = B^T V_2 \underset{\sim}{f}.$$

In (27),(28) the terms $0(r^3)$ represent a matrix and vector respectively, dependent on x and y with the indicated asymptotic properties in norm as $r^2 = x^2 + y^2 \to 0$ in any manner.

Equations (27) and (28) suggest the use of perturbation theory in the analysis of solutions of $A(z)\ \underset{\sim}{a}(z) = \underset{\sim}{\varphi}(z)$ and, consequently, of the local behaviour of the surface $f(z) = \underset{\sim}{b}^T(z)\ \underset{\sim}{a}(z)$ in a neighbourhood of the typical data point z_k.

We have seen that, at $z = z_k$ say, $A(z_k)$ is a matrix of rank one and so has one non-zero eigenvalue, say λ_1, and a zero eigenvalue of multiplicity $n-1$. Since $A(z_k) = \underset{\sim}{b}(z_k)\underset{\sim}{b}(z_k)^T$ it is clear that $\lambda_1 = \|\underset{\sim}{b}(z_k)\|^2$ and $\underset{\sim}{x}_1 = \|\underset{\sim}{b}(z_k)\|^{-1}\underset{\sim}{b}(z_k)$ is the associated eigenvector of unit length (we use the euclidean vector norm). For z in a sufficiently small deleted neighbourhood N_k of z_k and a well-posed problem, $\underset{\sim}{a}(z) = A(z)^{-1}\underset{\sim}{\varphi}(z)$, and the problem now is to analyze the behaviour of $A(z)^{-1}\underset{\sim}{\varphi}(z)$ as $z \to z_k$.

Let $\lambda_1(z),\ldots,\lambda_s(z)$ be the distinct eigenvalues of $A(z)$ with $\lambda_1(z) \to \lambda_1$ and $\lambda_r(z) \to 0$ $(r = 2,\ldots,s)$ as $z \to z_k$. Associate with each eigenvalue an orthogonal projector onto the associated eigenspace, $G_j(z)$, $j = 1,\ldots,s$. Then for $z \in N_k$,

$$A(z)^{-1} = \sum_{j=1}^{s} \frac{1}{\lambda_j(z)} G_j(z)\ . \tag{31}$$

In equation (27) let T be an orthogonal matrix for which $TA_0T^{-1} = \operatorname{diag}\{\lambda_1,0,\ldots,0\} =: D$. Then

$$TA(z)T^{-1} = D + r^2TA_2T^{-1} + O(r^3).$$

Applying Gersgorin's theorem to estimate the eigenvalues it is found that, as $r \to 0$

$$\begin{cases} \lambda_1(z) = \lambda_1 + r^2\lambda_1^{(2)} + O(r^3)\ , \\ \lambda_j(z) = r^2\lambda_j^{(2)} + O(r^3), \qquad j = 2,3,\ldots,s\ . \end{cases} \tag{32}$$

for some constants $\lambda_1^{(2)},\ldots,\lambda_s^{(2)}$.

It is known (ch. 5 of [3]) that if the fundamental Lagrange polynomials for $A(z)$ $(z \in N_k)$ are

$$\ell_k(\lambda) = \prod_{\substack{j=1\\ j\neq k}}^{s} (\lambda-\lambda_z)\Big/\prod_{\substack{j=1\\ j\neq k}}^{s} (\lambda_k-\lambda_j)$$

then $G_k(z) = \ell_k(A(z))$ and, given the result (32), it follows that the orthogonal projectors $G_k(z)$ have a similar asymptotic behaviour:

$$G_k(z) = G_k + r^2G_k^{(2)} + O(r^3) \tag{33}$$

as $r \to 0$ for $k = 1,2,\ldots,s$.

The following condition will be seen to have the effect of (a) restricting the order of singularity of $A^{-1}(z)$ at z_k (by ensuring that $\lambda_2^{(2)},\ldots,\lambda_s^{(2)}$ in (31) are non-zero) and (b) admitting a proof a C^1 smoothness for f in a neighbourhood of z_k .

DEFINITION The moving least squares method is *very well posed for* $\{z_j\}_{j=1}^N$ if (a) it is well-posed for every subset of $N-1$ points chosen from $\{z_j\}_{j=1}^N$ and (b) the condition (23) obtains.

LEMMA 2 *If the moving least squares method is very well posed then the coefficients* $\lambda_2^{(2)},\dots,\lambda_s^{(2)}$ *of* (32) *are positive.*

Proof Write $K=\sum_{j\neq k} d_{j,k}^{-2}$ for the kth diagonal element of V_0 in equation (26).

Observe that $B^T=\left[\underset{\sim}{b}(z_1)\dots\underset{\sim}{b}(z_N)\right]$ and so from (29):

$$A_2=B^TV_2B=\sum_{j\neq k} d_{j,k}^{-2}\underset{\sim}{b}(z_j)\underset{\sim}{b}^T(z_j)-K\underset{\sim}{b}(z_k)\underset{\sim}{b}^T(z_k)=A_k-KA_0$$

with the obvious definition of matrix A_k. Now on comparison with (12) it is clear that A_k has just the structure $A(z_k)$ except that the index k is missing from the summation. It follows that, with the very well posed hypothesis, A_k is positive definite (for any choice of k).

For $j=2,\dots,s$, the relation

$$G_j(z)A(z)G_j(z)=\lambda_j(z)G_j(z)$$

holds in N_k and using expansions of (27), (32) and (33) it is found that $G_jA_2G_j=\lambda_j^{(2)}G_j$; hence

$$\lambda_j^{(2)}G_j=G_j(A_k-KA_0)G_j,\qquad j=2,3,\dots,s.$$

Since $A_0=\lambda_1G_1, G_jA_0=0$ and $\lambda_j^{(2)}G_j=G_jA_kG_j$. The fact that A_k is positive definite then implies $\lambda_j^{(2)}>0$.//

The positivity of the $\lambda_j^{(2)}$ will play an important role in the theorem of section 5. However, it should be remarked that examples 1 and 2 demonstrate that the very well-posed hypothesis is not always necessary to guarantee a smooth surface $f(z)$. In those examples the problems are well-posed but not very well posed. In spite of this, the surfaces generated are C^∞.

5. THE SMOOTHNESS RESULT

THEOREM. *If the moving least squares method is very well posed and if the basis functions* $\{b_i(z)\}_{i=1}^N$ *are* C^1 *on* $\mathbb{R}^2$ *then the surface defined by the function* f *on* $\mathbb{R}^2\setminus\{z_j\}_{j=1}^N$ *and by continuation of* f, $\frac{\partial f}{\partial x}$ *and* $\frac{\partial f}{\partial y}$ *on the whole of* $\mathbb{R}^2$ *is a* C^1 *surface and interpolates the data.*

Proof Because of the representation (18) it is sufficient to prove the result for a cardinal function. In this case $f_j = \delta_{jk}$, $j = 1,\ldots,N$. In N_k (small enough) we have (see equations (16) and (21)),

$$\underset{\sim}{\varphi}(z) = B^T V \underset{\sim}{f} = v_k(z)\,\underset{\sim}{p}(z_k) = \lambda_1^{\frac{1}{2}} v_k(z) \underset{\sim}{x}_1$$

and using (24) with $K = \sum_{j \neq k} d_{j,k}^{-2}$,

$$\underset{\sim}{\varphi}(z) = \lambda_1^{\frac{1}{2}} \underset{\sim}{x}_1 (1 - Kr^2) + O(r^3), \quad \text{as } r \to 0 \,. \tag{34}$$

Then from (33), as $r \to 0$,

$$G_1(z)\underset{\sim}{\varphi}(z) = \lambda_1^{\frac{1}{2}} \left[I + r^2 (G_1^{(2)} - KG_1) \right] \underset{\sim}{x}_1 + O(r^3)$$

and for $j = 2,3,\ldots,s$,

$$G_j(z)\underset{\sim}{\varphi}(z) = \lambda_1^{\frac{1}{2}} r^2 G_j^{(2)} \underset{\sim}{x}_1 + O(r^3) \,.$$

Then using (31):

$$\underset{\sim}{a}(z) = A^{-1}(z)\underset{\sim}{\varphi}(z) = \lambda_1^{-1}(z) G_1(z) \underset{\sim}{\varphi}(z) + \sum_{j=2}^{s} \lambda_j^{-1}(z) G_j(z) \underset{\sim}{\varphi}(z)$$

$$= \lambda_1^{-\frac{1}{2}} \underset{\sim}{x}_1 + \sum_{j=2}^{s} \frac{\lambda_1^{\frac{1}{2}}}{\lambda_j^{(2)}} G_j^{(2)} \underset{\sim}{x}_1 + O(r)$$

and we have used the result of lemma 2. This proves the continuity of $\underset{\sim}{a}$, and hence f, at z_k.

The reference to the second order term $G_j^{(2)}$ can be removed from the limiting expression for $\underset{\sim}{a}(z_k)$ as follows: Substitute the "r-expansions" in the equation $G_j(z)A(z) = \lambda_j(z)G_j(z)$, $j = 2,\ldots,s$ and it is found that $G_j A_2 + G_j^{(2)} A_0 = G_j \lambda_j^{(2)}$. Since $A_0 = \lambda_1^{-1} G_1$ and $A_2 = A_k - KA_0$,

$$G_j A_k + \lambda_1^{-1} G_j^{(2)} G_1 = G_j \lambda_j^{(2)} . \tag{35}$$

Taking the composition with $\underset{\sim}{x}_1$, $\lambda_1^{-1} G_j^{(2)} \underset{\sim}{x}_1 = G_j A_k \underset{\sim}{x}_1$. It follows that

$$\underset{\sim}{a}(z_k) = \lambda_1^{-\frac{1}{2}} \left[I - \sum_{j=2}^{s} \frac{1}{\lambda_j^{(2)}} G_j A_k \right] \underset{\sim}{x}_1 \,. \tag{36}$$

The interpolation condition for the kth cardinal function at z_k is (cf. equation (16)),

$$\underset{\sim}{p}^T(z_k)\, \underset{\sim}{a}(z_k) = 1.$$

Since $\underset{\sim}{b}(z_k) = \lambda_1^{1/2}\underset{\sim}{x}_1$ and $\underset{\sim}{x}_1^T G_j = \underset{\sim}{0}^T$ for $j = 2,3,\ldots,s$, this obviously follows from (36).

To investigate the first derivative function $\frac{\partial f}{\partial x}$, we use relations (19) and (20). Using the expansions known for $\frac{\partial A}{\partial x}$, $\frac{\partial \underset{\sim}{\varphi}}{\partial x}$, and $\underset{\sim}{\alpha}$ ((27), (34) and (36) are relevant) it is found that in N_k,

$$(37) \qquad -\frac{\partial A}{\partial x}\underset{\sim}{\alpha} + \frac{\partial \underset{\sim}{\varphi}}{\partial x} = \frac{2x}{\lambda_1^{1/2}}\{A_k - KA_0 + KI - A_k\sum_{j=2}^{s}\frac{1}{\lambda_j^{(2)}}G_jA_k\}\underset{\sim}{x}_1 + 0(r^2).$$

Then from (27) and (32) it is clear that

$$\frac{\partial \underset{\sim}{\alpha}(z)}{\partial x} = A^{-1}(z)\left\{-\frac{\partial A}{\partial x}\underset{\sim}{\alpha} + \frac{\partial \underset{\sim}{\varphi}}{\partial x}\right\}$$

is continuous at z_k if and only if, for $i = 2,3,\ldots,s$,

$$G_i\left.\left(-\frac{\partial A}{\partial x}\underset{\sim}{\alpha} + \frac{\partial \underset{\sim}{\varphi}}{\partial x}\right)\right|_{z_k} = 0 .$$

Substituting from (37) and using the orthogonality conditions $G_iA_0 = 0$, $G_i\underset{\sim}{x}_1 = 0$, this is found to be equivalent to

$$(38) \qquad \left[G_i - \sum_{j=2}^{s}\frac{1}{\lambda_j^{(2)}}G_iA_kG_j\right]A_k\underset{\sim}{x}_1 = 0 , \qquad i = 2,\ldots,s .$$

But equation (35) implies that

$$G_iA_kG_j = \lambda_i^{(2)}\delta_{ij}G_i , \qquad i,j = 2,\ldots,s,$$

and hence, using lemma 2 again,

$$(39) \qquad \sum_{j=2}^{s}\frac{1}{\lambda_j^{(2)}}G_iA_kG_j = G_i , \qquad i = 2,\ldots,s.$$

Thus the condition (38) is satisfied and it is found that $\frac{\partial \underset{\sim}{\alpha}}{\partial x}$ is continuous in a neighbourhood of z_k. Continuity of $\frac{\partial \underset{\sim}{\alpha}}{\partial y}$ is established similarly.

It remains only to check the same properties for, say, the ℓth cardinal function $c_\ell(z)$ at the point z_k, $\ell \neq k$. Similar techniques apply. Instead of (34) we have

$$\underset{\sim}{\varphi}(z) = v_\ell(z)\underset{\sim}{b}(z_\ell) = r^2 d_{\ell,k}^{-2}\,\underset{\sim}{b}(z_\ell) + 0(r^3)$$

and from the equation $\underset{\sim}{\alpha}(z) = A^{-1}(z)\underset{\sim}{\varphi}(z)$ it is found that $\underset{\sim}{\alpha}$ is continuous at z_k if we define

$$(40) \qquad \underset{\sim}{\alpha}(z_k) = \left(\sum_{j=2}^{s}\frac{1}{\lambda_j^{(2)}}G_j\right)d_{\ell,k}^{-2}\,\underset{\sim}{b}(z_\ell) .$$

The interpolation condition is now

$$\underset{\sim}{k}^T(z_k)\ \underset{\sim}{a}\ (z_k) = 0$$

and this clearly follows from (40) and $\underset{\sim}{k}(z_k) = \lambda_1^{\frac{1}{2}} \underset{\sim}{x}_1$.

Then $\frac{\partial A}{\partial x} = 2x(A_k - \kappa A_0) + 0(r^2)$, and

$$\frac{\partial \underset{\sim}{\varphi}}{\partial x} = 2xd_{\ell,k}^{-2} \underset{\sim}{b}(z_\ell) + 0(r^2), \text{whence}$$

$$-\frac{\partial A}{\partial x} \underset{\sim}{a} + \frac{\partial \underset{\sim}{\varphi}}{\partial x} = 2x\left[-\sum_{j=2}^{s} \frac{1}{\lambda_j^{(2)}} A_k G_j + I\right] d_{\ell,k}^{-2} \underset{\sim}{b}(z_\ell) + 0(r^2).$$

Then it is found that $\frac{\partial \underset{\sim}{a}}{\partial x}$ and $\frac{\partial \underset{\sim}{a}}{\partial y}$ exist and are continuous only if

$$G_i = \sum_{j=2}^{s} \frac{1}{\lambda_j^{(2)}} G_i A_k G_j\ , \qquad i = 2,\ldots,s,$$

and this is just condition (39).//

ACKNOWLEDGEMENTS

The author is deeply indebted to his colleague, K. Salkauskas, for constant discussion and enthusiasm relating to this paper. In particular, he was the first to prove lemma 1. Susan Ritchie has provided computing assistance which is also gratefully acknowledged.

REFERENCES

[1] R.E. Barnhill, *Representation and approximation of surfaces*, Mathematical Software III. Academic Press, 1977, pp 69-120.

[2] W.J. Gordon and J.A. Wixom, *Shepard's method of "metric interpolation" to bivariate and multi-variate interpolation.* Math. Comp. 32 (1978), 253-264.

[3] P. Lancaster, *Theory of Matrices*, Academic Press, New York, 1969.

[4] D.H. McLain, *Drawing contours from arbitrary data points*, Computer J. 17 (1974), 318-324.

[5] D.H. McLain, *Two-dimensional interpolation from random data*, Computer J. 19 (1976), 178-181.

[6] S. Ritchie, *Representation of surfaces by finite elements*, M.Sc. Thesis, Dep't. of Math. & Stat., U. of Calgary, 1978.

[7] L.L. Schumaker, *Fitting surfaces to scattered data*, Approximation Theory, II, Academic Press, New York, 1976, pp. 203-268.

[8] D. Shepard, *A two-dimensional interpolation function for irregularly spaced data*, Proc. 1968 A.C.M. Nat. Conf., pp.517-524.

POLYNOMIAL SPLINES AND DIFFERENCE EQUATIONS

Günter Meinardus

University of Siegen, West Germany

Abstract: This is an elementary introduction to polynomial spline functions. The existence and uniqueness of the B-splines is proved. A special representation of those functions leads to many properties and, in particular, to the fact that they form a basis of the splines. The concept of linear difference equation is discussed. Eventually the connection of spline interpolation problems with linear difference equations is pointed out.

This lecture will give the simple connection between the concept of polynomial spline functions and that of linear difference equations. It therefore yields some explanation of the well-known fact that in interpolation problems with splines one has to deal with matrices of band structure.

1. POLYNOMIAL SPLINES.

We are given a bi-infinite sequence of real numbers x_r, $r \in \mathbb{Z}$, with

$$x_r < x_{r+1}, \quad r \in \mathbb{Z},$$

and

$$\lim_{r\to-\infty} x_r = -\infty, \quad \lim_{r\to\infty} x_r = +\infty.$$

The set of these numbers is referred to as the set K of knots.- Let furthermore m be a natural number, $m \geq 2$.

Definition: A real or complex function s on $\mathbb{R}$ is called a spline function of order m with respect to the set K, if $s \in C^{m-2}(\mathbb{R})$ holds, and if the restriction of s to the subinterval $I_r = [x_{r-1}, x_r]$ for all $r \in \mathbb{Z}$ belongs to the space π_{m-1} of polynomials of degree at most m-1.

Badri N. Sahney (ed.), Polynomial and Spline Approximation, 121-135.

Let p_r be the restriction of s to I_r, then there exists a number c_r such that

$$p_{r+1}(x)=p_r(x)+c_r(x-x_r)^{m-1}.$$

The set of all spline functions of order m with respect to the set K is a vector space over the complex field and will, in the following, be denoted by $S_m(K)$. The simplest spline functions in $S_m(K)$ are, for $r\in\mathbb{Z}$,

$$s_r(x) = \begin{cases} (x-x_r)^{m-1}, & \text{if } x\geq x_r, \\ 0 & \text{, if } x\leq x_r, \end{cases}$$

the so-called truncated power functions, for which we use the abbreviation $s_r(x) = (x-x_r)_+^{m-1}$.

The most important problem in vector spaces always is to find a suitable basis. The following finite support basis is due to H.B. CURRY and I.J. SCHOENBERG [2].

Theorem 1: For any given $r\in\mathbb{Z}$ there exists one and only one $s\in S_m(K)$ which satisfies

$$s(x) = 0 \ , \quad \text{if} \quad x\leq x_r, \tag{1}$$

$$s(x) = 0 \ , \quad \text{if} \quad x\geq x_{r+m}, \tag{2}$$

$$\int_{x_r}^{x_{r+m}} s(x)dx = 1. \tag{3}$$

Proof:

Any $s\in S_m(K)$ satisfying (1), (2), and (3) can be written as

$$s(x) = \sum_{k=0}^{m} b_k(x-x_{r+k})_+^{m-1}$$

with

$$\sum_{k=0}^{m} b_k(x-x_{r+k})^{m-1} = 0 \quad \text{for all} \quad x, \tag{4}$$

and

$$\sum_{k=0}^{m} b_k (x_{r+m}-x_{r+k})^m = m. \tag{5}$$

Comparing the coefficients of the powers of x in (4) and using (5) we are led to the linear system of equations for the numbers b_k,

$$\sum_{k=0}^{m} b_k (x_{r+k})^v = \begin{cases} 0, & \text{if } v=0,1,\ldots,m-1, \\ (-1)^m m, & \text{if } v=m, \end{cases}$$

which obviously has a unique solution.

The function $s \in S_m(K)$ satisfying (1), (2), and (3) is called the B-spline

$$B_m(x;x_r,x_{r+1},\ldots,x_{r+m}),$$

or, in a condensed form,

$$B_{mr}(x).$$

We now give a special representation of these functions. There are others, but the one given in the theorem below seems to be very comfortable for later use.

Theorem 2: Let x be the real variable and $\mathcal{C}_x$ a simply closed rectifiable curve in the complex z-plane such that all knots x_v with

$$x \leq x_v \leq x_{r+m}$$

and no others are in its interior. Integrating in the positive direction, the following formula holds:

$$B_{mr}(x) = \frac{1}{2\pi i} \int_{\mathcal{C}_x} \frac{m(z-x)^{m-1}\, dz}{(z-x_r)(z-x_{r+1})\ldots(z-x_{r+m})} \tag{6}$$

Proof:

In each of the intervals $I_w = [x_{w-1}, x_w]$ the right hand side of (6), which we denote for a moment by $\varphi_r(x)$, is a polynomial of degree at most $m-1$. If x moves from I_w to I_{w+1} the value of $\varphi_r(x)$ changes by an additional term of the form

$$c_w(x-x_w)^{m-1}$$

therefore $\varphi_r \in S_m(K)$. For $x<x_r$ the interior of $\mathcal{C}_x$ contains

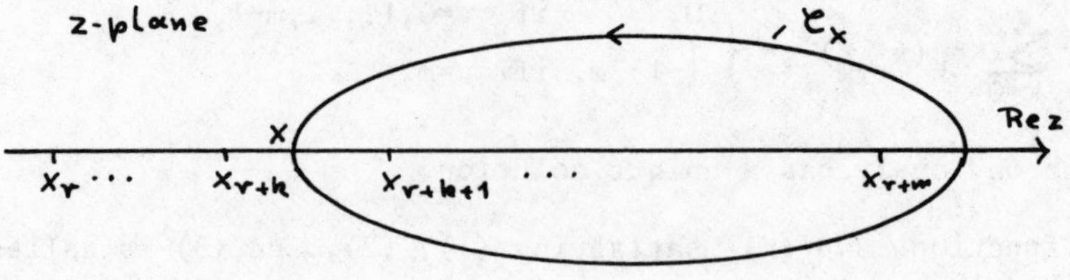

Fig. 1: Path of Integration in (6).

all the knots $x_r, x_{r+1}, \ldots, x_{r+m}$. Since the integrand is a rational function in z of degree $m-1$ in the numerator and of degree $m+1$ in the denominator, holomorphic in the exterior of $\mathcal{C}_x$, the integral vanishes. According to the Cauchy theorem we have likewise, that $\varphi_r(x)$ vanishes for $x>x_{r+m}$. In addition,

$$\int_{x_r}^{x_{r+m}} \varphi_r(x)dx = \frac{1}{2\pi i} \int_{\mathcal{C}_{x_r}} \frac{(z-x_r)^m dz}{(z-x_r)(z-x_{r+1})\ldots(z-x_{r+m})}$$

$$= 1.$$

The theorem 1 gives now the proposition (6) of theorem 2.

Theorem 3: For any $r\in\mathbb{Z}$ and $m\geq 3$ we have

$$\frac{d}{dx} B_{mr}(x) = m \frac{B_{m-1,r}(x)-B_{m-1,r+1}(x)}{x_{r+m}-x_r} . \quad (7)$$

Proof:

It is

$$\frac{d}{dx} B_{mr}(x) = -m \int_{\mathcal{C}_x} \frac{(m-1)(z-x)^{m-2} dz}{(z-x_r)\dots(z-x_{r+m})}$$

$$= \frac{-m}{x_{r+m}-x_r} \left\{ \frac{1}{2\pi i} \int_{\mathcal{C}_x} \frac{(m-1)(z-x)^{m-2} dz}{(z-x_{r+1})\dots(z-x_{r+m})} - \right.$$

$$\left. - \frac{1}{2\pi i} \int_{\mathcal{C}_x} \frac{(m-1)(z-x)^{m-2} dz}{(z-x_r)\dots(z-x_{r+m-1})} \right\}$$

$$= \frac{m}{x_{r+m}-x_r} \left\{ B_{m-1,r+1}(x) - B_{m-1,r}(x) \right\} .$$

Theorem 4: For any $r \in \mathbb{Z}$ we have

$$B_{mr}(x) = \left(\frac{m}{m-1}\right) \cdot \frac{\left(x-x_r\right) B_{m-1,r}(x) + \left(x_{r+m}-x\right) B_{m-1,r+1}(x)}{x_{r+m}-x_r} \quad (8).$$

Proof:

This follows from (6), using the decomposition

$$\frac{(z-x)^{m-1} \cdot (x_{r+m}-x_r)}{(z-x_r)\dots(z-x_{r+m})} = \frac{(x-x_r)(z-x)^{m-2}}{(z-x_r)\dots(z-x_{r+m-1})} - \frac{(x_{r+m}-x)(z-x)^{m-2}}{(z-x_{r+1})\dots(z-x_{r+m})} .$$

Corollary:

It is

$$B_{mr}(x) > 0$$

for all $x \in (x_r, x_{r+m})$.

Proof:

For $m=2$ one has

$$B_{2r}(x) = \begin{cases} 0 , & \text{if } x \le x_r \\ \dfrac{2(x-x_r)}{(x_{r+1}-x_r)(x_{r+2}-x_r)}, & \text{if } x_r \le x \le x_{r+1} \\ \dfrac{2(x_{r+2}-x)}{(x_{r+2}-x_{r+1})(x_{r+2}-x_r)}, & \text{if } x_{r+1} \le x \le x_{r+2} \\ 0 , & \text{if } x \ge x_{r+2} . \end{cases}$$

Therefore

$$B_{2r}(x) > 0 \quad \text{for} \quad x \in (x_r, x_{r+2}) .$$

This function $B_{2r}(x)$ is the so-called roof function.

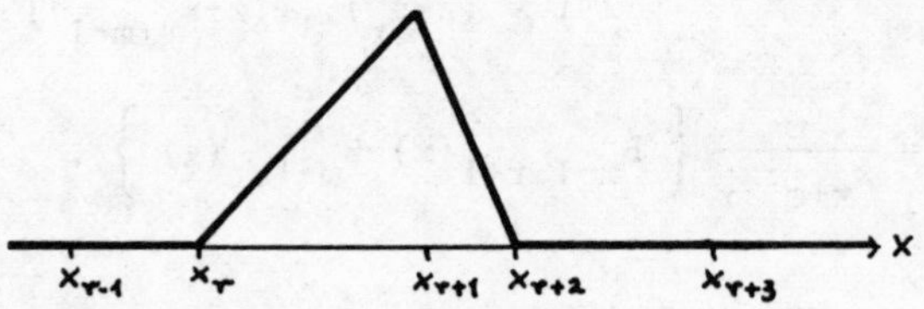

Fig. 2: Graph of $B_{2r}(x)$

Using (8), the proof of the corollary now follows by induction.

Because of the finite support of each B-spline, there is no convergence problem in considering series as

$$\sum_{r=-\infty}^{+\infty} a_r B_{mr}(x).$$

Theorem 5: For every $x \in \mathbb{R}$ we have the identity

$$\sum_{r=-\infty}^{+\infty} (x_{r+m} - x_r) B_{mr}(x) = m . \tag{9}$$

Proof:

We proceed by induction again. For $m=2$ and $x \in [x_\nu, x_{\nu+1}]$, $\nu \in \mathbb{Z}$, it is obvious that

$$\begin{aligned} \sum_{\nu=-\infty}^{+\infty} (x_{\nu+2} - x_\nu) B_{2\nu}(x) &= (x_{\nu+1} - x_{\nu-1}) B_{2,\nu-1}(x) + \\ &\quad + (x_{\nu+2} - x_\nu) B_{2\nu}(x) \\ &= 2 \end{aligned}$$

holds. The recursion formula (8) then leads to

$$\sum_{r=-\infty}^{+\infty} (x_{r+m}-x_r)B_{mr}(x) = \frac{m}{m-1}\left\{\sum_{r=-\infty}^{+\infty} (x-x_r)B_{m-1,r}(x) + \right.$$

$$\left. + \sum_{r=-\infty}^{+\infty} (x_{r+m}-x)B_{m-1,r+1}(x)\right\}$$

$$= \frac{m}{m-1}\cdot\sum_{r=-\infty}^{+\infty} (x_{r+m-1}-x_r)B_{m-1,r}(x)$$

$$= m \ .$$

Theorem 6: Every $s \in S_m(K)$ has a unique representation

$$s(x) = \sum_{r=-\infty}^{+\infty} a_r B_{mr}(x) \ .$$

Proof:

For $m=2$ and given $s \in S_2(K)$ let

$$a_r = \frac{1}{2} s(x_{r+1})(x_{r+2}-x_r) \ .$$

Then

$$\sum_{r=-\infty}^{+\infty} a_r B_{2r}(x)$$

coincides with s at the knots. Since the restriction of s to the interval I_r is a polynomial of degree at most 1. The same is true for the series. Therefore they are both uniquely determined by the value at the knots.

Now let $m \geq 3$. Again we proceed by induction. If $s \in S_m(K)$, then $s' \in S_{m-1}(K)$. Therefore s' has a unique representation

$$s'(x) = \sum_{r=-\infty}^{+\infty} b_r B_{m-1,r}(x) \ .$$

There exist numbers c_r, $r \in \mathbb{Z}$, such that

$$c_r - c_{r-1} = b_r \ .$$

This, by theorem 3, yields

$$s'(x) = \sum_{r=-\infty}^{+\infty} (c_r - c_{r-1}) B_{m-1,r}(x)$$

$$= \sum_{r=-\infty}^{+\infty} c_r (B_{m-1,r}(x) - B_{m-1,r+1}(x))$$

$$= \sum_{r=-\infty}^{+\infty} c_r \frac{(x_{r+m} - x_r)}{m} B'_{m\,r}(x) .$$

We therefore have

$$s(x) = \sum_{r=-\infty}^{+\infty} c_r \frac{(x_{r+m} - x_r)}{m} B_{mr}(x) + \gamma ,$$

where γ is a constant, which, according to (9) may be represented by

$$\gamma = \sum_{r=-\infty}^{+\infty} \gamma \frac{(x_{r+m} - x_r)}{m} B_{mr}(x) .$$

The existence of a representation of s is thus proved.

To show the uniqueness it is only needed to prove that the zero function has zero coefficients. From

$$\sum_{r=-\infty}^{+\infty} a_r B_{mr}(x) \equiv 0$$

follows

$$\sum_{r=-\infty}^{+\infty} a_r B'_{mr}(x) \equiv 0 .$$

Using theorem 3 again we arrive at

$$\sum_{r=-\infty}^{+\infty} \left\{ \frac{a_r}{x_{r+m} - x_r} - \frac{a_{r-1}}{x_{r+m-1} - x_{r-1}} \right\} B_{m-1,r}(x) \equiv 0 .$$

By induction it follows that

$$a_r = c(x_{r+m} - x_r)$$

with a constant c. But then

$$\sum_{r=-\infty}^{+\infty} a_r B_{mr}(x) = c \sum_{r=-\infty}^{+\infty} (x_{r+m}-x_r) B_{mr}(x)$$

$$= c \cdot m = 0 \ ,$$

therefore $c=0$ and $a_r=0$ for all $r \in \mathbb{Z}$.

The just proved basis property is important for the connection with difference equations.

We speak of the "equidistant case" in spline theory, if the knots are equally spaced, i.e. if there exists a real number $h>0$, such that

$$x_{r+1} = x_r + h \quad \text{for all} \quad r \in \mathbb{Z} \ ,$$

therefore

$$x_r = x_0 + rh \ , \quad r \in \mathbb{Z} \ .$$

We will give an explicit expression and some relations for the B-splines belonging to this set of equidistant knots. For simplicity we assume x_0 to be equal to zero and $h=1$.

Theorem 7: The B-splines for the equidistant case with $x_0=0$ and $h=1$ have the properties

1. $B_{mr}(x) = B_{m0}(x-r), \quad r \in \mathbb{Z}$, (10)

2. $B_{m,0}(x) = \int_{x-1}^{x} B_{m-1,0}(t)dt,$ (11)

3. $B_{m,0}(x) = \frac{1}{(m-1)!} \sum_{v=0}^{m} (-1)^v \binom{m}{v} (x-v)_+^{m-1}$ (12)

Proof:

The "shift"-property (10) follows at once from the defining properties of the B-splines, given in theorem 1. The recursive formula (11) is a consequence of theorem 3, since the derivative of the integral on the right hand side of (11) is equal to

$$B_{m-1,0}(x) - B_{m-1,1}(x) = B'_{m,0}(x) \ ,$$

according to (7), and the integral vanishes at $x=0$. The representation (12) eventually can be proved by induction, using (11): For $m=2$ we have

$$B_{20}(x) = (x)_+ - 2(x-1)_+ + (x-2)_+ =$$

$$= \begin{cases} 0 & , \text{ if } x \leq 0 , \\ x & , \text{ if } 0 \leq x \leq 1 , \\ 2-x & , \text{ if } 1 \leq x \leq 2 , \\ 0 & , \text{ if } 2 \leq x . \end{cases}$$

Therefore, for $m \geq 3$,

$$B_{m,0}(x) = \frac{1}{(m-2)!} \cdot \sum_{v=2}^{m-1} (-1)^v \binom{m-1}{v} \cdot \int_{x-1}^{x} (t-v)_+^{m-2} dt$$

$$= \frac{1}{(m-1)!} \sum_{v=0}^{m} (-1)^v \binom{m}{v} (x-v)_+^{m-1}$$

after a little algebra.

2. LINEAR DIFFERENCE EQUATIONS.

We consider maps

$$\varphi : \mathbb{Z} \to \mathbb{C} .$$

Any such map is called a bi-infinite sequence of complex numbers. The vector space of those sequences over the complex field is referred to as the sequence space Q. The elements of Q are sometimes denoted by

$$a = \{a_\nu\}_{\nu \in \mathbb{Z}} , \; b \in \{b_\nu\}_{\nu \in \mathbb{Z}} , \text{ etc.}$$

If we are given sequences

$$a^{(0)}, a^{(1)}, \ldots , a^{(k)} \text{ and } b$$

from Q, we call the infinite system of equations

$$a_r^{(0)} y_{r+k} + a_r^{(1)} y_{r+k-1} + \ldots + a_r^{(k)} y_r = b_r , \quad r \in \mathbb{Z} , \tag{13}$$

a linear difference equation of order k for the sequence $y \in Q$. Furthermore we speak of a regular linear difference equation, if all the numbers $a_r^{(0)}, a_r^{(k)}$, $r \in \mathbb{Z}$, are different from zero.

It is easy to prove that there exist sequences

$$y^{(1)}, y^{(2)}, \ldots , y^{(k)} ,$$

linearly independent in Q, such that the linear space of solutions of the regular homogeneous linear difference equation (13), i.e. $b=0$, is spanned by these k sequences. If $y^{(0)}$ is a special solution of the inhomogeneous equation, then the general solution consists of the addition of this sequence $y^{(0)}$ and the general solution of the homogeneous equation. All this is very similar to the analogue case of an ordinary linear differential equation.

In the following we assume that all the numbers $a_r^{(0)}$ as well as the numbers $a_r^{(k)}$ of the given linear difference equation (13) are different from zero.

An initial value problem for (13) would be: We look for a sequence $y \in Q$, satisfying (13), such that the k numbers

$$y_0, y_1, \ldots, y_{k-1}$$

of the sequence y are prescribed ("initial values"). According to our assumption it is obvious that there exists one and only one y which satisfies all the conditions.

We give three examples of boundary value problems for the difference equation (13):

In a two-point linear boundary value problem we are given a natural number (mostly with $n \geq k$) and k linearly independent constraints, either of the form

$$\sum_{r=0}^{k-2} \alpha_{1r} y_r = \gamma_1 ,$$

or of the form

$$\sum_{r=0}^{k-2} \beta_{1r} y_{r+n} = \delta_1 .$$

A typical example is, for $k=2m$,

$$y_1 = \gamma_1 ; \; y_{1+n} = \delta_1 ; \quad 1 = 0,1,\ldots,m-1 .$$

Again one asks for a sequence $y \in Q$ satisfying both, the constraints and the equation (13).

The periodic boundary value is as follows: For a given natural number n the difference equation (13) has "periodic coefficients", i.e. the numbers

$$a_r^{(0)}, a_r^{(1)}, \ldots, a_r^{(b)}, b_r$$

depend only on the residue class of r modulo n :

$$a_{r_1}^{(\mu)} = a_{r_2}^{(\mu)} \text{ and } b_{r_1} = b_{r_2}$$

whenever

$$r_1 \equiv r_2 \bmod n ,$$

$\mu = 0,1,\ldots,n$. Then we ask for a "periodic" solution $y \in Q$ of (13), i.e. such a y, for which

$$y_{r_1} = y_{r_2}$$

whenever $r_1 \equiv r_2 \bmod n$. The periodicity of y is equivalent to the boundary conditions

$$\begin{aligned} y_0 &= y_n , \\ y_1 &= y_{n+1} , \\ &\ldots \\ y_{k-1} &= y_{n+k-1} , \end{aligned}$$

because of the periodicity of the coefficients.

As boundary conditions at infinity one frequently considers growth constraints on the sequence y. One asks, for example, for such solutions $y \in Q$ of (13) for which real numbers c and λ exist such that

$$|y_r| \leqq c(|r|+1)^{\lambda}$$

holds for all $r \in \mathbb{Z}$.

3. SPLINE INTERPOLATION.

We state a class of spline interpolation problems: Let τ be a fixed real number

$$0 < \tau \leqq 1 .$$

We are looking for a spline function $s \in S_m(K)$ which interpolates a given real or complex function f at the points

$$x_{r-1} + \tau(x_r - x_{r-1}) ,$$

either for all $r \in \mathbb{Z}$ or for $r=0,1,\ldots,n$ with a given natural number n :

$$s(x_{r-1}+\tau(x_r-x_{r-1})) = f(x_{r-1}+\tau(x_r-x_{r-1})) = b_r .$$

This problem does not have, in general, a unique solution.

According to theorem 6 we may use the B-spline basis to derive an equivalent problem in the sequence space Q. If

$$0<\tau<1$$

we put

$$s(x) = \sum_{\nu=-\infty}^{+\infty} y_{\nu+m} B_{m\nu}(x) .$$

The interpolation condition yields

$$s(x_{r-1}+\tau(x_r-x_{r-1})) = \sum_{\nu=r-m}^{r-1} y_{\nu+m} B_{m\nu}(x_{r-1}+\tau(x_r-x_{r-1})) = b_r ,$$

or

$$a_r^{(0)} y_{r+m-1} + a_r^{(1)} y_{r+m-2} + \ldots + a_r^{(m-1)} y_r = b_r$$

with

$$a_r^{(0)} = B_{m,r-m}(x_{r-1}+\tau(x_r-x_{r-1})) ,$$

$$a_r^{(1)} = B_{m,r-m+1}(x_{r-1}+\tau(x_r-x_{r-1})) ,$$

$$\cdots\cdots\cdots\cdots\cdots\cdots$$

$$a_r^{(m-1)} = B_{m,r-1}(x_{r-1}+\tau(x_r-x_{r-1})) .$$

All these numbers are positive. We therefore get a regular linear difference equation of order $m-1$.

Let $\tau=1$. This means interpolation at the knots x_r :

$$s(x_r) = f(x_r) = c_r .$$

In this case, putting

$$s(x) = \sum_{\nu=-\infty}^{+\infty} y_{\nu+m-1} B_{m\nu}(x) ,$$

we have

$$s(x_r) = \sum_{\nu=r-m+1}^{r-1} y_{\nu+m-1} B_{m\nu}(x_r) = c_r ,$$

or

$$d_r^{(0)} y_{r+m-2} + d_r^{(1)} y_{r+m-3} + \ldots + d_r^{(m-2)} y_r = c_r$$

with

$$d_r^{(0)} = B_{m,r-m+1}(x_r) ,$$

$$d_r^{(1)} = B_{m,r-m+2}(x_r) ,$$

$$\ldots\ldots\ldots\ldots\ldots\ldots\ldots$$

$$d_r^{(m-2)} = B_{m,r-1}(x_r) ,$$

which again are all positive numbers. This time we are led to a regular linear difference equation of order $m-2$.

In the equidistant case (without restriction with $x_0=0$, $h=1$, i.e. $x_r=r$), we have, in the first case, $0<\tau<1$, for $\mu=0,1,\ldots,m-1$:

$$a_r^{(\mu)} = B_{m,r-m+\mu}(r-1+\tau)$$
$$= B_{m,0}(m-\mu-1+\tau) ,$$

independent of r.

In the second case, $\tau=1$, we have for $\mu=0,1,\ldots,m-2$:

$$d_r^{(\mu)} = B_{m,r-m+\mu+1}(r)$$
$$= B_{m,0}(m-\mu-1) ,$$

again independent of r.

We conclude that in the equidistant case our spline interpolation problem leads to constant coefficients in the homogeneous part of the corresponding difference equation. We remark furthermore that, for an arbitrary set of knots K the interpolation problems yields the lowest order of the corresponding difference equation (namely $m-2$), if the interpolation at the knots is concerned.

There are many different ways of selecting a unique solution of this interpolation problem, using a boundary value principle. We will specialize to periodic splines now, where we prefer to use a vectorial notation.

[1] C. DE BOOR: On Calculating with B-Splines
J. Approximation Theory 6 (1972), 5o-62.

[2] H.B. CURRY and I.J. SCHOENBERG: On Polya Frequency Functions IV: The fundamental Spline Functions and their

Limits. J. Anal. Math. 17 (1966), 71-107.

[3] G. MEINARDUS: Bemerkungen zur Theorie der B-Splines. Symp. Public.: Spline-Funktionen, edit. by K. Böhmer, G. Meinardus, W. Schempp. Bibl. Institut Mannheim (1974), 165-175.

PERIODIC SPLINES

Günter Meinardus

University of Siegen, West Germany

Abstract: Periodic spline vectors, gained by a simple transformation from periodic splines, are introduced. Furthermore some interpolation problems are discussed. The equidistant case, leading to circulant matrices, yields explicit representation of the interpolating vector. This Lagrange type formula is discussed to some extent.

Let n be a natural number. The set K of knots x_r from now should have a special "periodic" structure: If $r_1, r_2 \in \mathbb{Z}$ and

$$r_1 = r_2 + \lambda n, \quad \lambda \in \mathbb{Z},$$

then

$$x_{r_1} = x_{r_2} + \lambda (x_n - x_o).$$

If one knows the knots $x_o, x_1, \ldots, x_n$, then the set K may be constructed by periodic continuation of these knots, the period being $x_n - x_o$.

A periodic spline function s of order m belonging to the set K is an element of $S_m(K)$, which is periodic with period $x_n - x_o$. Let $p_r(x)$ be the restriction of such function s to the interval $[x_{r-1}, x_r]$. Then there exist numbers c_r, depending only on the residue class of r modulo n, such that

$$p_{r+1}(x) = p_r(x) + c_r (x - x_r)^{m-1} \tag{1}$$

and

$$p_{n+r}(x) = p_r(x - (x_n - x_o)) \tag{2}$$

for $r \in Z$.

We consider, for $t \in [0,1]$, polynomials $q_r(t)$, which are defined by

$$q_r(t) = p_r(x_{r-1} + t(x_r - x_{r-1})). \tag{3}$$

Badri N. Sahney (ed.), Polynomial and Spline Approximation, 137-146.

Using the notation

$$\alpha_r = \frac{x_{r+1}-x_r}{x_r-x_{r-1}} \tag{4}$$

we conclude

$$\alpha_{r_1}=\alpha_{r_2} \tag{5}$$

and, from (2),

$$q_{r_1}(t)=q_{r_2}(t) \tag{6}$$

for $r_1 \equiv r_2 \pmod n$. Furthermore, the relation (1) yields

$$q_{r+1}(t)=q_r(1+\alpha_r t)+\tilde{c}_r t^{m-1} \tag{7}$$

for $r \in Z$, where

$$\tilde{c}_r=c_r(x_{r+1}-x_r)^{m-1} .$$

It therefore seems natural to consider the column vector

$$q(t)=\begin{pmatrix} q_1(t) \\ q_2(t) \\ \vdots \\ q_n(t) \end{pmatrix}$$

with components belonging to the space $\pi_{m-1}[0,1]$. We call such vector a periodic spline vector of order m, belonging to

$$\alpha=\begin{pmatrix} \alpha_1 \\ \alpha_2 \\ \vdots \\ \alpha_n \end{pmatrix} ,$$

if it satisfies (7) for $r=1,2,\ldots,n$ with

$$q_{n+1}=q_1 .$$

The vector space of these vectors is denoted by $P_m(\alpha)$. From (4) it follows that the positive numbers α_r are restricted to the condition

$$\alpha_1\alpha_2\ldots\alpha_n=1 . \tag{8}$$

The equidistant case is equivalent to

$$\alpha_1=\alpha_2=\ldots=\alpha_n=1.$$

In the following we use the matrix

$$T = \begin{pmatrix} 0 & \cdots & & 0 & 1 \\ 1 & \ddots & & & 0 \\ 0 & 1 & \ddots & & \vdots \\ \vdots & & \ddots & \ddots & \vdots \\ 0 & \cdots & 0 & 1 & 0 \end{pmatrix}$$

which permutes cyclically the components of a n-dimensional vector:

$$w = \begin{pmatrix} w_1 \\ w_2 \\ \cdot \\ \cdot \\ \cdot \\ w_n \end{pmatrix}, \quad Tw = \begin{pmatrix} w_n \\ w_1 \\ w_2 \\ \vdots \\ w_{n-1} \end{pmatrix}$$

Theorem 1: Let $m<n$. Then there exists a spline vector of order m,

$$q(t,\alpha) = \begin{pmatrix} q_1(t,\alpha) \\ q_2(t,\alpha) \\ \cdot \\ \cdot \\ \cdot \\ q_n(t,\alpha) \end{pmatrix},$$

which has the following properties:

1. For $r=m+1,m+2,\ldots,n$ we have

$$q_r(t,\alpha)=0. \tag{9}$$

2. The polynomials

$$q_2(t,\alpha);q_3(t,\alpha);\ldots;q_{m-1}(t,\alpha)$$

are positive for $t\in[0,1]$.

3. There exist positive constants $a_m(\alpha),b_m(\alpha)$, such that

$$q_1(t,\alpha)=a_m(\alpha)\, t^{m-1} \tag{10}$$

and

$$q_m(t,\alpha)=b_m(\alpha)(1-t)^{m-1}. \tag{11}$$

4. The vector is uniquely defined up to a positive factor.

5. The spline vectors

$$T^{\nu} q(t, T^{-\nu}\alpha) \;;\quad \nu=0,1,\ldots,n-1; \tag{12}$$

form a basis of the vector space $P_m(\alpha)$.

Proof: By theorem 6 of the foregoing lecture we know of the basis property of the B-splines. If K has the periodic structure we see that, for

$$r_1 = r_2 + \lambda n ,$$

we have the relation

$$B_{mr_2}(x) = B_{mr_1}(x + \lambda(x_n - x_o)).$$

If furthermore

$$s(x) = \sum_{r=-\infty}^{+\infty} d_r B_{mr}(x)$$

is periodic, then

$$s(x) = s(x + x_n - x_o) = \sum_{r=-\infty}^{+\infty} d_r B_{mr}(x + x_n - x_o)$$

$$= \sum_{r=-\infty}^{+\infty} d_r B_{m,r+n}(x)$$

$$= \sum_{r=-\infty}^{+\infty} d_{r-n} B_{mr}(x) ,$$

therefore

$$d_{r_1} = d_{r_2}$$

whenever $r_1 \equiv r_2 \pmod{n}$.

The n series

$$\sum_{\lambda=-\infty}^{+} B_{m,\nu+\lambda n}(x) \quad \nu=0,1,\ldots,n-1 \;; \tag{13}$$

consequently form a basis of the vector space of periodic splines of order m, belonging to the set K.

Consider now in (13) the case $\nu=0$. Our transformation (3) to the spline vectors yields a vector $q(t,\alpha)$ which has the properties 1,2,3 and 4. It is easy to prove that the transformation (3) of (13) for the subscript ν leads to the spline vector

$$T^{\nu}q(t,T^{-\nu}\alpha)\ ;\ \nu=0,1,\dots,n-1 .$$

Corollary: For any admissible vector α we have

$$\dim P_m(\alpha)=n . \tag{14}$$

The simplest result in spline interpolation can be obtained for an odd number n.

Let, for odd n, be y a given vector,

$$y = \begin{pmatrix} y_1 \\ y_2 \\ \vdots \\ y_n \end{pmatrix},\ y_{\nu}\in\mathbb{C} .$$

We want to have a spline vector $q\in P_m(\alpha)$, which, for given $\tau\in(0,1]$, has the property

$$q(\tau)=y . \tag{15}$$

This interpolation problem is equivalent to the one stated in the foregoing paper, if, in addition, the set K and the spline function s are periodic.

Theorem 2: There is, for odd n, one and only one $q\in P_m(\alpha)$ which satisfies (15).

Proof: We go back to the spline functions $s(x)$. The existence and uniqueness of s, interpolating y in the following way

$$s(x_{r-1}+\tau(x_r-x_{r-1}))=y_r ;\ r=1,2,\dots,n ;$$

is proved, if one proves that the homogeneous problem $y=0$ has only the trivial solution. By way of contradiction we assume that there exists a spline function s, not vanishing identically, such that

$$s(x_{r-1}+\tau(x_r-x_{r-1}))=0;\ r=1,2,\ldots,n\ .$$

If there is a subscript ν, such that

$$s(x)=0 \text{ for all } x\in[x_{\nu-1},x_\nu]\ ,$$

then we have the contradiction $s(x)=0$ everywhere, using (1) and the interpolatory properties. Therefore s has the n isolated zeros $x_{r-1}+\tau(x_r-x_{r-1})$. We now refer to a special case of a theorem due to L. SCHUMAKER [3], which says that the number of isolated zeros of a periodic spline function in the interval $(x_o,x_n]$ is at most equal to $n-1$, if n is odd and if s is not identically zero.

With the basis of $P_m(\alpha)$, given by

$$T^\nu q(t,T^{-\nu}\alpha);\ \nu=0,1,\ldots,n-1;$$

we form the basis matrix

$$W_m(t,\alpha) = ((q(t,\alpha),Tq(t,T^{-1}\alpha),\ldots,T^{n-1}q(t,T^{-n+1}\alpha)),$$

in which we assume that the normalization

$$q_1(t,\alpha)=t^{m-1} \tag{16}$$

has been used. Then the main diagonal of $W_m(t,\alpha)$ consists only of the powers t^{m-1}. The theorem 2 means, for odd n, that $W_m(\tau,\alpha)$ is nonsingular. Any $q\in P_m(\alpha)$ can be written as

$$q=W_m b\ ,$$

with a suitable vector $b\in\mathbb{C}^n$. The interpolation problem is solved if the inverse of $W_m(\tau,\alpha)$ has been constructed.

We give, in the equidistant case, where we just write $W_m(t)$, some special results for these matrices.

Theorem 3: The matrix $W_m(t)$ is uniquely defined by the properties

$$W_m^{(\mu)}(0)=TW_m^{(\mu)}(1);\mu=0,1,\ldots,m-2; \tag{17}$$

where the superscript μ denotes the derivative of μth order.

Proof: The first column $q(t)$ of $W_m(t)$ has the form

$$q(t) = \begin{pmatrix} t^{m-1} \\ q_2(t) \\ \vdots \\ q_m(t) \\ 0 \\ \vdots \\ 0 \end{pmatrix}$$

with

$$q_{\nu+1}(t)=q_\nu(1+t)+d_\nu t^{m-1} \quad .$$

Therefore

$$q(t)=Tq(1+t)+dt^{m-1} \tag{18}$$

with a constant vector d. It follows

$$q^{(\mu)}(0)=Tq^{(\mu)}(1) \text{ for } \mu=0,1,\ldots,m-2 \; .$$

The other columns of $W_m(t)$ are just the permutations of q :

$$T^\nu q \; ; \; \nu=1,2,\ldots,n-1 \; .$$

Remark: We have

$$W_2(t) = tI+(1-t)T$$

and

$$W_3(t) = t^2I+(1+2t-2t^2)T+(1-t)^2T^2 \; .$$

Theorem 4: For $m=2,3,\ldots$ the following recursion formula holds:

$$W_{m+1}(t)= \int_o^t W_m(z)dz+T \int_t^1 W_m(z)dz \; . \tag{19}$$

The proof follows easily from theorem 3.

Definition: Let $M_n(t)$ be the set of all $n\times n$ matrices whose entries are polynomials over $\mathbb{C}$. A matrix $B\in M_n(t)$ is called a circulant if it commutes with T:

$$TB=BT \; .$$

Theorem 5: Every circulant matrix B is a polynomial in T, i.e. there exist polynomials $b_o(t)$, $b_1(t),\ldots$, $b_{n-1}(t)$, such that

$$B=b_o(t)I+b_1(t)T+\ldots+b_{n-1}(t)T^{n-1} \; .$$

Proof: Every matrix $B\in M_n(t)$ can be represented uniquely in the form

$$B=D_o(t)+D_1(t)T+\dots+D_{n-1}(t)T^{n-1}$$

with diagonal matrices $D_\nu(t)$. The condition

$$TB=BT$$

leads to

$$TD_\nu(t)=D_\nu(t)T\ ;\quad \nu=0,1,\dots,n-1\ .$$

This is only possible if

$$D_\nu(t)=b_\nu(t)I\ ;\quad \nu=0,1,\dots,n-1\ ,$$

with a polynomial $b_\nu(t)$.

Definition: Let

$$\zeta = \exp\left(\frac{2\pi i}{n}\right)$$

a primitive root of unity. Then the matrix

$$U_n = \frac{1}{\sqrt{n}}\,((\zeta^{\nu\mu}))\ ;\ \nu,\mu=0,1,\dots,n-1\ ; \tag{20}$$

is called the matrix of the discrete FOURIER transform.

The following properties of U_n are easily verified:

1. U_n is an unitary matrix:

$$U_n^*=U_n^{-1} \tag{21}$$

2. $$U_n^4=I \tag{22}$$

3. $U_n^2=P$ is the permutation matrix:

$$P=((p_{\nu\mu}))\,;\nu,\mu=0,1,\dots,n-1\ ;$$

$$p_{\nu\mu}=\begin{cases} 1\ , & \text{if } \nu+\mu\equiv 0 \pmod n \\ 0\ , & \text{if } \nu+\mu\not\equiv 0 \pmod n\ . \end{cases}$$

4. $$U_n T U_n^*=Z \tag{23}$$

with $$Z=\mathrm{diag}(1,\zeta,\zeta^2,\dots,\zeta^{n-1})\ .$$

Theorem 6: Every circulant matrix B can be transformed to diagonal form, using the unitary matrix U_n:

$$U_n B U_n^*=\mathrm{diag}\ .$$

Theorem 7: The matrices $W_m(t)$ are circulant.

Proof: The matrix

$$TW_m(t)T^{-1}$$

satisfies all the conditions of theorem 3 and therefore coincides with $W_m(t)$.

Definition: Let, for $t\in[0,1]$ and z a complex variable,

$$H_o(t,z)\equiv 1,\ H_1(t,z)=z+(1-z)t,$$

$$H_{r+1}(t,z)=\{(1-z)t+(r+1)z\}H_r(t,z)+z(1-z)\frac{\partial}{\partial z}H_r(t,z)\ ,$$

for $r=0,1,2,\ldots$.

Theorem 8: The polynomials $H_r(t,z)$ satisfy the functional equation

$$H_r(t,z)-zH_r(1+t,z)=(1-z)^{r+1}t^r \tag{24}$$

The proof can be found in [2] .
The interpolation problem with n odd,

$$q\in P_m\ ,\ q(\tau)=y$$

has the explicit solution

$$q(t)=W_m(t)W_m^{-1}(\tau)y\ . \tag{25}$$

The matrix

$$W_m(t)W_m^{-1}(\tau)\ , \tag{26}$$

which corresponds to a LAGRANGE type representation of the interpolation problem, is a circulant matrix. It can be characterized as follows:

Theorem 9: Let

$$Q_m(t)=\mathrm{diag}\ H_{m-1}(t,\zeta^{\mu})\ ;\ \mu=0,1,\ldots,n-1\ .$$

Then

$$W_m(t)W_m^{-1}(\tau)=U_n^*Q_m(t)Q_m^{-1}(\tau)U_n. \tag{27}$$

Proof: For $t=\tau$ both sides of (27) coincide with the unit matrix. Using (18), we get the functional equation

$$W_m(t)W_m^{-1}(\tau)=TW_m(1+t)W_m^{-1}(\tau)+Ft^{m-1} ,$$

F being a constant matrix. This equation together with the normalization defines the matrix uniquely.

The functional equation (24) shows that the right hand side of (27) has the same property.

Literatur.

[1] J.H. AHLBERG, E.N. NILSON and J.L. WALSH:
The theory of splines and their applications.
Academic Press, New York (1967).

[2] G. MEINARDUS and G. MERZ: Periodische Splinefunktionen.
In: Spline-Funktionen, Bibliogr. Inst. Mannheim.
Edited by K. Böhmer, G. Meinardus, W. Schempp, 177-195 (1974).

[3] L. SCHUMAKER: Zeros of Spline Functions and Applications.
Journ. Appr. Th., 18, 152-168 (1976).

PERIODIC SPLINES AND FOURIER ANALYSIS

Günter Meinardus

University of Siegen, West Germany

Abstract: The so-called discrete Fourier transform is introduced. Then the connection with Fourier coefficients of 2π-periodic functions is described. A theorem for some linear mapping leads to the theory of attenuation factors due to L. Collatz and W. Quade. Finally a new version of the Fast Fourier transform is presented.

Let f(t) be a periodic function of the real variable t with period 2π. For Riemann-integrable f we have for the coefficients a_k of the FOURIER series

$$\sum_{k=-\infty}^{+\infty} a_k e^{ikt}$$

the representation

$$a_k = \frac{1}{2\pi}\int_0^{2\pi} f(t)e^{-ikt}dt, \; k\in\mathbb{Z}.$$

The simplest quadrature formulas for these integrals are

$$c_k = \frac{1}{n}\sum_{\lambda=0}^{n-1} y_\lambda \zeta^{-\lambda k} \tag{1}$$

with

$$y_\lambda = \frac{1}{2\pi} f\left(\frac{2\pi\lambda}{n}\right); \; n\in\mathbb{N} .$$

The formula (1) may be interpreted as derived from an interpolation problem for polynomials: We want to construct a polynomial $p\in\pi_{n-1}$,

$$p(z) = \sum_{k=0}^{n-1} c_k z^k \tag{2}$$

which solves the interpolation problem

Badri N. Sahney (ed.), Polynomial and Spline Approximation, 147–154.

$$p(\zeta^{\lambda}) = y_{\lambda} \; ; \quad \lambda=0,1,\dots,n-1. \tag{3}$$

Furthermore, the vectors

$$c = \begin{pmatrix} c_o \\ c_1 \\ \cdot \\ \vdots \\ c_{n-1} \end{pmatrix} \quad \text{and} \quad y = \begin{pmatrix} y_o \\ y_1 \\ \cdot \\ \vdots \\ y_{n-1} \end{pmatrix}$$

satisfy

$$c = \frac{1}{\sqrt{n}} PU_n y , \tag{4}$$

where U_n is the unitary matrix from the foregoing lecture, and $P \stackrel{\text{n}}{=} U_n^2$.

A further connection with the FOURIER coefficients is contained in the

Theorem 1: Let $g(z)$ be a function of the complex variable z, holomorphic in the annulus

$$r<|z|<R$$

with $0<r<1<R$.

Let

$$y_{\lambda}=g(\zeta^{\lambda}) \; ; \quad \lambda=0,1,\dots,n-1.$$

Then the coefficients c_k in (2) are given by the series

$$c_k = \sum_{r=-\infty}^{+\infty} a_{k+rn} \; ; \; k=0,1,\dots,n-1. \tag{5}$$

Remark: The coefficients a_k are just the FOURIER coefficients of the 2π-periodic function

$$f(t)=g(e^{it}) .$$

In order to consider other approximations to the FOURIER coefficients than the one given by (1) we use the following concept: We denote by $P_{2\pi}$ the linear space of 2π-periodic complex-valued functions of the variable t. Let us consider now maps

$$\varphi:\mathbb{C}^n \to P_{2\pi} .$$

A vector $y\in\mathbb{C}^n$ will be written as

$$y = \begin{pmatrix} y_o \\ y_1 \\ \vdots \\ y_{n-1} \end{pmatrix} .$$

One example is the map which associates to y the function $p(e^{it})$, where $p(z)$ is the interpolating polynomial of (2), (3).

In general we write

$$\varphi(y)=f \ , \ f\in P_{2\pi} \ . \tag{6}$$

Such a map φ is called invariant under translation, if from (6) we always may conclude

$$\varphi(Ty)=\tilde{f} \ , \tag{7}$$

where

$$\tilde{f}(t)=f(t-\frac{2\pi}{n}) \ . \tag{8}$$

Here T denotes the permutation matrix already used in the foregoing lecture.
Let L be a linear functional on $P_{2\pi}$ with the property, that there exists a complex number $\beta=\beta_L$ such that

$$L(\tilde{f})=\beta L(f) \tag{9}$$

for all $f\in P_{2\pi}$. In the following we use the notation for a basis of $\mathbb{C}^n$:

$$e_o = \begin{pmatrix} 1 \\ 0 \\ \vdots \\ 0 \end{pmatrix} , \ e_1 = \begin{pmatrix} 0 \\ 1 \\ 0 \\ \vdots \\ 0 \end{pmatrix} = Te_o \ , \ \dots \ , e_{n-1} = \begin{pmatrix} 0 \\ \vdots \\ 0 \\ 1 \end{pmatrix} = T^{n-1} e_o \ .$$

Theorem 2: If the map

$$\varphi : \mathbb{C}^n \to P_{2\pi}$$

is linear and invariant under translation, and if the linear functional L on $P_{2\pi}$ satisfies (9), then we have

$$L(\varphi(y))=L(\varphi(e_o)) \cdot \sum_{\nu=o}^{n-1} y_\nu \beta^\nu \ . \tag{10}$$

Proof: According to our assumptions we have

$$\varphi(y) = \sum_{\nu=o}^{n-1} y_\nu \varphi(e_\nu) = \sum_{\nu=o}^{n-1} y_\nu \varphi(T^\nu e_o) = \sum_{\nu=o}^{n-1} y_\nu g_\nu$$

with

$$g_\nu \in P_{2\pi}$$

and

$$g_o = \varphi(e_o) \ , \ g_{\nu+1} = \tilde{g}_\nu \ , \ \nu = 0,1,\ldots,n-2 \ .$$

Therefore, because of (9),

$$L(\varphi(y)) = \sum_{\nu=o}^{n-1} y_\nu L(g_\nu) = \sum_{\nu=o}^{n-1} y_\nu \beta^\nu \cdot L(g_o) \ .$$

In our context the functional L stands for the FOURIER coefficients:

$$L(f) = L_\lambda(f) = \frac{1}{2\pi} \int_o^{2\pi} f(t) e^{-i\lambda t} dt \ .$$

We have

$$L_\lambda(\tilde{f}) = \frac{1}{2\pi} \int_o^{2\pi} f(t - \frac{2\pi}{n}) e^{-i\lambda t} dt$$

$$= e^{-\frac{2\pi i\lambda}{n}} \cdot \frac{1}{2\pi} \int_o^{2\pi} f(t) e^{-i\lambda t} dt$$

therefore

$$\beta = \beta(\lambda) = e^{-\frac{2\pi i\lambda}{n}} = \zeta^{-\lambda} \ .$$

The special mapping, defined by (2), (3) yields

$$L_\lambda(\varphi(y)) = c_\lambda = L_\lambda(\varphi(e_o)) \cdot \sum_{\nu=o}^{n-1} y_\nu \zeta^{-\lambda\nu} \ .$$

Obviously

$$\varphi(e_o) = \frac{1}{n} p_o(e^{it}) \ ,$$

with

$$p_o(z) = 1 + z + \ldots + z^{n-1} \ .$$

It follows

$$L_\lambda(\varphi(e_o)) = \frac{1}{n} \ ,$$

independent of λ.

The factor

$$L_\lambda(\varphi(e_o))$$

for general φ is called an attenuation factor. It occurred in a mathematical theory the first time in the famous paper by L. Collatz and W. Quade [1] on interpolation of periodic functions. Another paper on this topic is due to W. Gautschi [4]. In [1] one may find the first full treatment of periodic splines with equidistant knots. If the periodic function f for which we like to construct approximations of its FOURIER coefficients, belongs to the differentiability class C^r, r>0, but not to the class C^{r+1}, then it seems to be reasonable to use approximations for f by periodic spline interpolation of order r+2. If we are only interested in the FOURIER coefficients of this approximating function, we may apply the theorem 2. This means only to compute the interpolating spline function of the vector e.

Let us close this lecture with a new version of the Fast Fourier Transform (FFT)-method to carry out the interpolation problem (2), (3) in a most efficient way. The first step in this field is due to J.W. COOLEY and J.W. TUKEY [2]. A survey on this method with applications can be found in [6]. The new version, contained in [5], is as follows:

Let $n=2^r$, $r\in\mathbb{N}$ and, again,

$$\zeta = \exp\left(\frac{2\pi i}{n}\right).$$

Definition: For all pairs (ν,μ) with

$$\mu=0,1,\ldots,r;\ \nu=0,1,\ldots,2^{r-\mu}-1$$

let $p_{\nu,\mu}$ be the polynomial of degree at most $2^\mu-1$ which solves the interpolation problem:

$$p_{\nu,\mu}(\zeta^{\nu+\lambda\cdot 2^{r-\mu}})=y_{\nu+\lambda\cdot 2^{r-\mu}};\lambda=0,1,\ldots,2^\mu-1.$$

Remark: According to this definition we have

$$p_{\nu,0}\equiv y_\nu \quad \text{for} \quad \nu=0,1,\ldots,2^r-1$$

and

$$p_{0,r}=p\ ,$$

where p is defined by (2),(3).

Theorem 3: For $\mu = 0,1,\ldots,r-1;\ \nu = 0,1,\ldots,2^{r-\mu}-1$ we have the recursion formula:

$$p_{\nu,\mu+1}(z) = \frac{1}{2}\left(1+\zeta^{-\nu 2^{\mu}}\cdot z^{2^{\mu}}\right)\cdot p_{\nu,\mu}(z)+ \tag{11}$$
$$+\frac{1}{2}\left(1-\zeta^{-\nu 2^{\mu}}\cdot z^{2^{\mu}}\right)\cdot p_{\nu+2^{r-\mu-1},\mu}(z) .$$

Proof: We have to show that the right hand side of (11) satisfies all the conditions which define $p_{\nu,\mu+1}$ uniquely. Obviously, it is a polynomial of degree at most $2^{\mu+1}-1$. For

$$z = \zeta^{\nu+\lambda\cdot 2^{r-\mu-1}}$$

we have

$$\zeta^{-\nu\cdot 2^{\mu}} z^{2^{\mu}} = \zeta^{\lambda\cdot 2^{r-1}} = \exp(\pi i\lambda) = \begin{cases} 1 & \text{for even } \lambda, \\ -1 & \text{for odd } \lambda . \end{cases}$$

Hence, for even λ the second term on the right hand side of (11) vanishes, the first term is equal to

$$p_{\nu,\mu}\left(\zeta^{\nu+\frac{\lambda}{2}\cdot 2^{r-\mu}}\right) = y_{\nu+\lambda\cdot 2^{r-\mu-1}} .$$

The analogue holds for odd values of λ.
Another proof of (11) may be of interest: Let ρ be a real number, $\rho>1$. Then, integrating in the positive direction, the following contour integral representation is valid:

$$p_{\nu,\mu}(z) = \frac{1}{2\pi i}\int_{|w|=\rho}\left(\frac{w^{2^{\mu}}-z^{2^{\mu}}}{w-z}\right)\frac{p(w)\,dw}{(w^{2^{\mu}}-\zeta^{\nu 2^{\mu}})} .$$

Therefore

$$p_{\nu+2^{r-\mu-1},\mu}(z) = \frac{1}{2\pi i}\int_{|w|=\rho}\left(\frac{w^{2^{\mu}}-z^{2^{\mu}}}{w-z}\right)\frac{p(w)\,dw}{(w^{2^{\mu}}+\zeta^{\nu 2^{\mu}})} .$$

The combination of these two polynomials as given in (11) yield at once the desired result.

The numerical implementation of the recursion formula (11) can be done as follows: Let

$$p_{\nu,\mu}(z) = \sum_{\lambda=0}^{2^{\mu}-1} a_{\nu,\mu,\lambda} z^{\lambda} .$$

Then

$$a_{\nu,0,0} = y_{\nu} \; ; \quad \nu = 0,1,\ldots,2^{r}-1.$$

By (11) it is

$$\frac{1}{2}(a_{\nu,\mu,\lambda}+a_{\nu+2^{r-\mu-1},\mu,\lambda}) = a_{\nu,\mu+1,\lambda}$$

and

$$\frac{1}{2}\zeta^{-\nu\cdot 2^{\mu}}(a_{\nu,\mu,\lambda}-a_{\nu+2^{r-\mu-1},\mu,\lambda}) = a_{\nu,\mu+1,\lambda+2^{\mu}} ,$$

both for $\lambda = 0,1,\dots,2^{\mu}-1$. One may arrange the numbers $a_{\nu,\mu,\lambda}$ in a scheme:

$p_{\nu,0}$	$p_{\nu,1}$	$p_{\nu,2}$	$\cdots$
$a_{0,0,0}$	$a_{0,1,0}$	$a_{0,2,0}$	$\cdots$
$a_{1,0,0}$	$a_{0,1,1}$	$a_{0,2,1}$	$\cdots$
$a_{2,0,0}$	$a_{1,1,0}$	$a_{0,2,2}$	$\cdots$
$a_{3,0,0}$	$a_{1,1,1}$	$a_{0,2,3}$	$\cdots$
$a_{4,0,0}$	$a_{2,1,0}$	$a_{1,2,0}$	$\cdots$
$\vdots$	$\vdots$	$\vdots$	

It is obvious that the number of multiplications and additions in this procedure is proportional to $r\cdot 2^{r}$.

References:

[1] COLLATZ, L., QUADE, W.: Zur Interpolationstheorie der reellen periodischen Funktionen. Sitzungsber. Preuss. Akad. Wiss. 30,383-429(1938).

[2] COOLEY, J.W., TUKEY, J.W.: An algorithm for the machine calculation of complex Fourier series. Math. Comp. 19,297-301(1965).

[3] EHLICH, H.: Untersuchungen zur numerischen Fourieranalyse. Math. Z. 91,380-420 (1966).

[4] GAUTSCHI, W.: Attenuation Factors in Practical Fourier Analysis. Numer. Math. 18,

373-400(1972).

[5] MEINARDUS, G.: Schnelle Fourier-Transformation. In: Numerische Methoden der Approximationstheorie Bd. 4. ISNM 42, edited by L. Collatz, G. Meinardus, H. Werner. Birkhäuser-Verlag Basel (1978).

[6] SCHÜSSLER, H.W.: Digitale Systeme zur Signalverarbeitung. Springer-Verlag Berlin (1973).

COMPUTATION OF THE NORMS OF SOME SPLINE INTERPOLATION OPERATORS

Günter Meinardus

University of Siegen, West Germany

Abstract: This is a survey of investigations concerning norm computation, respectively norm estimations, of some spline interpolation operators. Problems of boundedness with respect to the number of knots, minimal norm problems and eventually explicit expression in the equidistant case are discussed.

Let $m \geqq 2$ be the order of the periodic spline functions s belonging to the set

$$K = \{x_\nu\}_0^n$$

of knots. Furthermore let τ be the real number with

$$0 < \tau \leqq 1,$$

with which we consider the interpolating problem

$$s(x_{\nu-1} + \tau(x_\nu - x_{\nu-1})) = y_\nu; \quad \nu = 1, 2, \ldots, n.$$

If this problem has a solution for every vector with components $y_\nu \in \mathbb{C}$, then the interpolating spline function is always unique and is the value of a bounded linear operator S, defined on $\mathbb{C}^n$. This operator depends on m, n, K and τ. We therefore write

$$S(m,n,K,\tau).$$

In this lecture we will give some results on investigations about bounds for S. For simplicity and because of the intimate connection with stability problems we will consider bounds based on norms. In addition we will restrict ourselves to a special case:

Definition: Let, for functions,

$$\|s\| = \sup_x |s(x)|$$

Badri N. Sahney (ed.), Polynomial and Spline Approximation, 155-161.

and, for vectors $y \in \mathbb{C}^n$,

$$||y|| = \max_{\nu} |y_\nu|.$$

Then we denote

$$R(m,n,K,\tau) = \sup_{\substack{y \in \mathbb{C}^n \\ ||y|| \leq 1}} ||S(m,n,K,\tau)y|| . \tag{1}$$

In the equidistant case (without restriction we may assume $x_\nu = \nu$) we suppress the letter K and write simply $R(m,n,\tau)$. In the case $\tau=1$ (interpolating at the knots) we suppress the letter τ and write simply $R(m,n,K)$, etc.

In the terminology of the second lecture we denote

$$\rho = \operatorname{Max}_{\nu} \operatorname{Max} (\alpha_\nu, 1/\alpha_\nu) \tag{2}$$

In [2] E.W. CHENEY and F. SCHURER proved several results on upper and lower bounds in the cubic case, i.e. $m = 4$. Let us mention a few of those: If

$$v = \operatorname{Max}_{\nu} \left(\frac{1}{h_\nu (1+\alpha_\nu)} \operatorname{Max} (\alpha_\nu, 1/\alpha_\nu)\right)$$

and

$$h = \operatorname{Max}_{\nu} h_\nu \quad , \quad h_\nu = x_\nu - x_{\nu-1} \; ,$$

then

$$1 \leq R(4,n,K) \leq \frac{3}{3} vh + 1 . \tag{3}$$

Another lower bound is

$$R(4,n,K) \geq \frac{\sqrt{3}}{36} \rho - 1 . \tag{4}$$

The method is based on a linear system of equations for the derivatives

$$\lambda_\nu = s'(x_\nu)$$

of the interpolating cubic spline

$$\frac{1}{1+\alpha_\nu} \lambda_{\nu+1} + 2\lambda_\nu + \frac{\alpha_\nu}{1+\alpha_\nu} \lambda_{\nu-1} =$$

$$= \frac{3}{h_\nu} \left\{ \frac{1}{\alpha_\nu (1+\alpha_\nu)} y_{\nu+1} - \frac{1-\alpha_\nu}{\alpha_\nu} y_\nu + \frac{\alpha_\nu}{1+\alpha_\nu} y_{\nu-1} \right\} ;$$

$\nu = 1,2,\ldots,n$,

and the fact that a cubic spline function can be computed (and hence estimated) by its values and its derivatives at the knots via Hermite interpolation.

There are some very interesting questions with respect to our operator norms:

1. If n is fixed, what kind of distribution of the knots yields the smallest norm? It is likely (cp [3]) that the equidistant case gives the smallest norm. To prove this seems to be extremely difficult. There is only one simple example, for which this conjecture has been proved [10]: Let us consider the quadratic case $m=3$ and n odd. Then the operator $S(3,n,K)$ exists and the following theorem is true

Theorem 1 (cp [10]): It is

$$R(3,n,K) = \frac{1}{2}\left\{1 + h\sum_{\nu=1}^{n}\frac{1}{h_\nu}\right\} \tag{5}$$

with

$$h = \max_\nu h_\nu .$$

Corollary : We have

$$R(3,n,K) \geqq R(3,n) = \frac{n+1}{2} \tag{6}$$

with equality only in the equidistant case.

The proof of the corollary is obvious.

2. In [4] it is proved that in the equidistant cubic case the norms are bounded. As a matter of fact, the following formula is valid:

$$R(4,n) = \frac{1}{4}\left\{1 + 3\sqrt{3}\,\frac{1-(2-\sqrt{3})^\lambda}{1+(2-\sqrt{3})^\lambda}\right\}, \tag{7}$$

where

$$\lambda = \begin{cases} n, & \text{if } n \text{ is odd,} \\ \frac{n}{2}, & \text{if } n \text{ is even.} \end{cases}$$

The difference equation for the derivatives of the interpolating spline function in this case has, because of $\alpha_\nu = 1$, constant coefficients. This gives the possibility to prove (7) by elementary methods. One observes the bound

$$R(4,n) \leqq \frac{1+3\sqrt{3}}{4} \approx 1.549038\ldots .$$

It is not difficult to prove that the recursion formula

$$R(4,2r+3) = \frac{13+14R(4,2r+1)}{10+8R(4,2r+1)}$$

is valid for $r = 0,1,2 \ldots$. With

$$R(4,1) = 1$$

this gives the result that these norms are rational numbers which can be computed easily:

$$R(4,3) = \frac{3}{2},\ R(4,5) = \frac{127}{82},\ R(4,7) = \frac{79}{51},\ \text{etc.} -$$

The question arises: For which values of the number ρ, defined in (2), one has also a bound for $R(4,n,K)$, independent of n ? The answer, in historical order, is:

For all

$$\rho < \sqrt{2} \tag{8}$$

(AMEIR and A.SHARMA [11]),

$$\rho < 2 \tag{9}$$

(E.W. CHENEY and F.SCHURER [3]),

$$\rho < 1 + \sqrt{2} \tag{10}$$

(C.A.HALL [6]),

$$\rho < 2.43916207 \ldots = \rho_0 , \tag{11}$$

ρ_0 = the only real zero of the polynomial

$$2t^3-t^2-7t-6$$

(M.MARSDEN [7]). In [7] it is proved, that for $\rho > \frac{3+\sqrt{5}}{2} \approx$ $\approx 2.61803398 \ldots$ there exists always a sequence of sets of knots $K = K_n$ such that the corresponding sequence of norms is unbounded. This is true for $\rho = \frac{3+\sqrt{5}}{2}$, too. The conjecture, stated in [7], that $\frac{3+\sqrt{5}}{2}$ is the best constant has been proved independently by C.DE BOOR [1] and by the author [8]. The boundedness follows from

Theorem 2 (cp. [8]): For general distribution of knots and with defined by (2) the estimation

$$R(4,n,K) \leq 1 + \frac{(4\rho-1)\rho^4}{2\lambda(\rho)} \cdot \frac{v_n^2(\rho)}{(1-2^{-n})^2} \tag{12}$$

holds. Here $\lambda(\rho)$ stands for

$$\lambda(\rho) = 1 + \rho + \sqrt{1+\rho+\rho^2}$$

and

$$v_n(\rho) = \begin{cases} \dfrac{1-(\rho^2/\lambda(\rho))^n}{1-\rho^2/\lambda(\rho)} & \text{for } \rho \geqq 1, \rho \neq \dfrac{3+\sqrt{5}}{2} \\ n & \text{for } \rho = \dfrac{3+\sqrt{5}}{2}. \end{cases}$$

Corollary: The norm $R(4,n,K)$ is bounded by the number

$$1 + \frac{2(4\rho-1)\rho^4}{\lambda(\rho)(1-\rho^2/\lambda(\rho))^2}$$

for all ρ with

$$\rho < \frac{3+\sqrt{5}}{2}.$$

The proof of (12) is complicated and is based on a factorization principle of some matrix.

3. In [12] and [9] the norms for the equidistant case but for arbitrary even order $m = 2k+2$ is computed. Using the explicit formula for a matrix representation of the interpolation operator (cp. theorem 9 of the second lecture) it is possible to prove some formulas for the norm (cp. [9]). For simplicity we restrict ourselves to the case where n is odd, $n = 2r+1$. The other cases can be and have been handled analoguously.

Theorem 3 (cp. [9]): For $r,k = 0,1,2,\ldots$ the following formulas are valid:

$$R(2k+2,2r+1) = \frac{1}{2r+1} \sum_{\mu=0}^{2r} \frac{2H_{2k+1}(\frac{1}{2},\zeta^\mu)}{(1+\zeta^\mu)H_{2k+1}(1,\zeta^\mu)}, \tag{13}$$

$$= \frac{1}{\pi i} \int_C \left(\frac{1+z^{2r+1}}{1-z^{2r+1}}\right) \frac{H_{2k+1}(\frac{1}{2},z)dz}{z(1+z)H_{2k+1}(1,z)}, \tag{14}$$

where

$$\zeta = \exp\left(\frac{2\pi i}{2r+1}\right).$$

The contour C is a circle

$$|z| = 1-\varepsilon,$$

ε being positive and small enough that all the k zeros of $H_{2k+1}(1,z)$ inside the unit circle are in the interior of C. Integration in (14) is done in the positive direction.

Remarks: Since the rational function

$$\frac{2H_{2k+1}(\frac{1}{2},z)}{(1+z)H_{2k+1}(1,z)}$$

has rational coefficients, the sum (13) belongs to the cyclotomic field generated by ζ. If one replaces ζ by any other $(2r+1)^{st}$ primitive root of unity the sum (13) remains unchanged. Hence all the algebraic conjugates of the algebraic number, defined by (13), are identical; therefore all the norms are given by rational numbers.

From the representation (14) we are able, by means of the residue theorem, to give explicit expressions of the norms in terms of the zeros of $H_{2k+1}(1,z)$ inside the unit circle. These zeros are all real and negative. The type of the formulas we get is as follows:

$$R(2k+2,2r+1) = a_0 + \sum_{\nu=1}^{k} a_\nu \frac{1-\rho_\nu^{2r+1}}{1+\rho_\nu^{2r+1}} \tag{15}$$

with real and positive numbers $a_0,a_1,\ldots,a_k$, and with real and positive numbers ρ_ν less than one. The formula (7) is the simplest example (k=1).

For other spaces e.g. L_p $(1\leq p<\infty)$ there are many investigations and interesting results. We refer to the literature for this.

References:

[1] DE BOOR, C.: On cubic spline functions which vanish at all knots. MRC Report No. 1424 (1974).

[2] CHENEY, E.W. and F.SCHURER: A note on the operators arising in spline approximation. JAT 1, 94-102 (1968).

[3] CHENEY, E.W. and F.SCHURER: Convergence of cubic spline interpolants. J.A.T. 3, 114-116 (1970).

[4] CHENEY, E.W. and F.SCHURER: On interpolating cubic splines with equally spaced nodes. Indag. Math. 30, 517-524 (1968).

[5] GOLOMB, M.: Approximation by periodic splines on uniform meshes. J.A.T. 1, 26-65 (1968).

[6] HALL, C.A.: Uniform Convergence of Cubic Spline Interpolants. J.A.T. 7,71-75 (1973)

[7] MARSDEN, M.: Cubic spline interpolation of continous functions. J.A.T. 10, 103-111 (1974)

[8] MEINARDUS, G.: Periodische Splinefunktionen. In: Spline Functions. Lecture Notes in Mathematics, Vol. 501, 177-199, Springer-Verlag (1975)

[9] MEINARDUS,G. und G.MERZ: Zur periodischen Spline-Interpolation. Erschienen in: Spline-Funktionen, Hrsg. K. Böhmer, G. Meinardus und W. Schempp. BI-Verlag Mannheim (1974)

[10] MEINARDUS,G. and G.D.TAYLOR: Periodic Quadratic Spline Interpolant of Minimal Norm. J.A.T. 23, 137-141 (1978)

[11] MEIR,A. and A.SHARMA: On Uniform Approximation by Cubic Splines. J.A.T. 2, 270-274 (1969)

[12] RICHARDS, F.B.: Best bounds for the uniform periodic spline interpolation operator. J.A.T. 7, 302-317 (1973)

AN INTRINSIC APPROACH TO MULTIVARIATE SPLINE INTERPOLATION AT ARBITRARY POINTS

Jean Meinguet

Université Catholique de Louvain

Institut de Mathématique Pure et Appliquée

chemin du Cyclotron 2

B-1348 Louvain-la-Neuve (Belgium)

ABSTRACT

The mathematical theory underlying the practical method of "surface spline interpolation" provides approximation theory with an intrinsic concept of multivariate spline ready for use. A proper abstract setting is shown to be some Hilbert function space, the reproducing kernel of which involves no functions more complicated than logarithms. The crux of the matter is that convenient representation formulas can be obtained by resorting to convolutions or to Fourier transforms of distributions.

1. THE OPTIMAL INTERPOLATION PROBLEM

Except in Section 2.2, all functions and vector spaces considered in this paper are real.

The classical problem of minimizing the quadratic functional

$$(1) \qquad |v|_m^2 := \int_a^b |v^{(m)}(t)|^2 \, dt$$

Badri N. Sahney (ed.), Polynomial and Spline Approximation, 163–190.

under the interpolatory constraints

(2a) $v(a_i) = \alpha_i, \quad 1 \leqslant i \leqslant N,$

with $N \geqslant m \geqslant 1$ and

(2b) $-\infty < a \leqslant a_1 < a_2 < \ldots < a_N \leqslant b < +\infty,$

has for natural setting the vector space

(3) $$H^m[a,b] := \{v \in C^{m-1}[a,b] : v^{(m-1)} \text{ absolutely continuous, } v^{(m)} \in L_2(a,b)\},$$

or equivalently, the class of functions whose distributional derivatives up to the m-th one are in $L_2(a,b)$. Equipped with the Sobolev seminorm $|\cdot|_m$, $H^m[a,b]$ is a semi-Hilbert function space, the linear variety defined by (2) being thus closed and Hausdorff. The existence of a unique solution u of the above problem, and its fundamental characterization as an odd degree polynomial spline, readily follow from the orthogonal projection theorem; a further analysis of this close association between spline interpolation and orthogonal projection leads to the so-called first integral relation (see e.g. [1] ,p.155), from which various important intrinsic properties (such as, for example, the minimum norm property and the best approximation property) can be obtained easily.

Multivariate splines are most often constructed from one-dimensional ones, by tensor product or blending techniques; however convenient this approach may prove occasionally, it suffers from the two inherent shortcomings: restriction to rectangular-like regions and meshes, use of somewhat unnatural function spaces and norms. These shortcomings are adequately covered by the intrinsically multivariate approach analyzed in the present paper; as a matter of fact, we will be concerned here mainly with the *mathematical foundations of surface spline interpolation* (for a more concrete presentation, stressing for example the significant properties of this optimal interpolation process at the algorithmic level, see [12]). Originally introduced by engineers (for interpolating wing deflections and computing slopes for aeroelastic calculations, see [10]), the ingenious mathematical tool called "surface spline" is simply a plate of infinite extent that deforms in bending only, its deflections being specified at a finite number of independent points (under the action of point loads to be determined accordingly). As regards the underlying mathematical analysis, many deep results have been obtained recently by Duchon (see [5,6,7]); while relying of course on that work, our approach is deliberately more constructive, a prominent role being played indeed by *representation formulas* in function and distribution spaces (see specially Section 2.1).

It turns out that a proper abstract setting for the present n-dimensional generalization of the above (univariate) problem and results is provided by the (generalized) *Beppo Levi space* $X \equiv BL^m(\mathbb{R}^n)$ of order m over $\mathbb{R}^n$ (m and n are integers $\geqslant 1$, to be regarded throughout as given). Defined by

$$(4) \qquad X := \{v \in \mathcal{D}' : \partial^\alpha v \in L_2 \quad \text{for } |\alpha| = m\},$$

where $\alpha := (\alpha_1,\ldots,\alpha_n) \in \mathbb{N}^n$ and $|\alpha| := \alpha_1 + \ldots + \alpha_n$, X is thus simply the vector space of all the (Schwartz) distributions (i.e., continuous linear functionals on the vector space $\mathcal{D}$ of infinitely differentiable functions with compact support in $\mathbb{R}^n$, provided with the canonical Schwartz topology) for which all the partial derivatives (in the distributional sense) of (total) order m are square integrable in $\mathbb{R}^n$ (for the Lebesgue measure dx).

X is naturally equipped with the semi-inner product $(.,.)_m$ corresponding to the seminorm

(5) $$|v|_m := |D^m v|_0 \equiv \{ \sum_{i_1,\ldots,i_m=1}^{n} \int_{\mathbb{R}^n} |\partial_{i_1\ldots i_m} v(x)|^2 dx\}^{1/2},$$

where $D^m v$, the m-th total derivative of v on $\mathbb{R}^n$, denotes the (m-fold , completely symmetric) tensor-valued function whose coordinates with respect to the canonical basis of $\mathbb{R}^n$ are the usual partial derivatives $\partial_{i_1\ldots i_m} v$ (to be interpreted in the distributional sense) for $i_1,\ldots,i_m \in [1,n]$; by a classical theorem (see e.g. [14], p.60), the kernel of $|\,.\,|_m$ is simply the vector space $\mathcal{P} \equiv \mathcal{P}_{m-1}$ of dimension

(6) $$M = \binom{n+m-1}{n}$$

of all polynomials over $\mathbb{R}^n$ of (total) degree $\leqslant m - 1$. It should be noticed that the variant (5) of the usual Sobolev m-seminorm is both translation and *rotation invariant*, while its homogeneity trivially reflects the action of dilations on $\mathbb{R}^n$; it is therefore to be expected that Fourier transforms will play an important role in the sequel (see Section 2.2). Moreover, just as the quadratic functional (1) can be physically interpreted (if m = 2) as the potential energy of a statically deflected thin beam (which indeed is proportional to the integral of the square of the curvature of the elastica of the beam), so its multivariate generalization defined by (5) can be regarded (if m = n = 2 and under some additional conditions) as the *bending energy of a thin plate* of infinite extent. As a matter of fact, it might seem here more natural to replace $\mathbb{R}^n$ by some compact subset; however, as already emphasized by Duchon, such an alternative possibility (considered for example by Atteia, see e.g. [2])

proves much more complicated to study and to exploit : it is indeed closely connected with highly non-trivial boundary value problems.

Beppo Levi spaces have many interesting properties (see specially [4]), which are partly reminiscent of those of the widely known *Sobolev spaces* (on $\mathbb{R}^n$)

(7) $$H^m := \{v \in L_2 : \partial^\alpha v \in L_2 \quad \text{for } |\alpha| \leqslant m\};$$

equipped with the inner product $((.,.))_m$ corresponding to the (rotation invariant) norm

(8) $$\| v \|_m = \Big(\sum_{j=0}^{m} | v |_j^2 \Big)^{1/2},$$

H^m is a Hilbert space. As regards Beppo Levi spaces, the following result is the only one, however, to be directly relevant here.

Theorem 1. *Suppose*

(9) $$m > n/2.$$

Then the seminormed space $X \equiv BL^m(\mathbb{R}^n)$, *defined by* (4) *and* (5), *is a semi-Hilbert function space of continuous functions on* $\mathbb{R}^n$, *all the evaluation linear functionals with finite support in* $\mathbb{R}^n$ *that annihilate* $P \equiv P_{m-1}$ *being accordingly bounded.*

Unlike Duchon in [6], where this basic result is presented as a straightforward consequence of an "iterated" version of a theorem essentially due to Krylov (see e.g. [14], p. 181 and p. 188), we will proceed here more constructively, by establishing in Section 2.1 representation formulas in X.

Let us come now to the precise statement of the optimal interpolation problem we want to analyze here, namely Problem (P). Let there be given :

- a finite set $A = (a_i)_{i \in I}$ of distinct points of $\mathbb{R}^n$ containing a *$\mathcal{P}$-unisolvent* subset, by which we mean a set $B = (a_j)_{j \in J}$ of M points of A, M being defined by (6), such that there exists a unique $p \in \mathcal{P}$ satisfying the interpolating conditions

$$(10) \qquad p(a_j) = \alpha_j, \quad \forall\, j \in J,$$

for any prescribed real scalars α_j, $\forall\, j \in J$.

- a set of real scalars $(\alpha_i)_{i \in I}$, or equivalently provided that $m > n/2$, the linear variety (of codimension Card(A)) defined by

$$(11) \qquad V := \{v \in X : v(a_i) = \alpha_i, \ \forall\, i \in I\};$$

whenever $\alpha_i = f(a_i)$, $\forall\, i \in I$, where f denotes a function defined on A, V can be interpreted as the set of X-*interpolants* of f on A.

Then we have the following definition of *Problem* (P) : *Find* $u \in V$ *such that*

$$(12) \qquad |u|_m = \inf_{v \in V} |v|_m,$$

it being understood once and for all that $m > n/2$.

2. REPRESENTATION FORMULAS IN BEPPO LEVI SPACES

2.1. A Convolution Approach

An eventually appropriate way of expressing a distribution $v \in \mathcal{D}'$ in terms of its partial derivatives $\partial_{i_1 \dots i_m} v$ of a given order $m \geqslant 1$ is suggested by the (partly formal) identities

$$(13)\qquad v = \delta * v = \Delta^m E * v = \sum_{i_1,\ldots,i_m=1}^{n} \partial_{i_1\ldots i_m} E * \partial_{i_1\ldots i_m} v,$$

which stem from the following facts (classical in the theory of distributions) :

- the n-dimensional Dirac distribution (or measure) δ is the unit of convolution : in symbols, $\delta * v = v$, $\forall\, v \in \mathcal{D}'$; more generally, the convolution of the shifted Dirac distribution $\delta_{(a)}$ with any distribution v yields the transform of v by the translation $x \mapsto x + a$ of $\mathbb{R}^n$.
- there exists a distribution E satisfying the partial differential equation

$$(14)\qquad \Delta^m E \equiv \sum_{i_1,\ldots,i_m=1}^{n} (\partial_{i_1\ldots i_m})^2 E = \delta;$$

such a *fundamental solution* of the *iterated Laplacian* Δ^m in $\mathbb{R}^n$ is the rotation invariant function on $\mathbb{R}^n - \{0\}$ defined by the following formulas (see e.g. [14], p. 288) :

$$(15)\qquad E(x) \equiv E_{m,n}(x) := \begin{cases} c\, r^{2m-n} \ln r, & \text{if } 2m \geqslant n \text{ and } n \text{ is even,} \\ d\, r^{2m-n} & \text{otherwise,} \end{cases}$$

where $r(x) \equiv |x|$ denotes as usual the radial coordinate (or Euclidean norm) of the point $x \in \mathbb{R}^n$ and

$$(16a)\qquad c \equiv c_{m,n} := \frac{(-1)^{n/2+1}}{2^{2m-1}\pi^{n/2}(m-1)!(m-n/2)!},$$

$$(16b)\qquad d \equiv d_{m,n} := \frac{(-1)^m\Gamma(n/2-m)}{2^{2m}\pi^{n/2}(m-1)!}.$$

- the product of *convolution* $v * w$ *of distributions* v and w, which is the distribution uniquely defined (if existent at all !) by the equation

(17) $$< v * w, \varphi > := < v_\xi, < w_\eta, \varphi(\xi+\eta) \gg, \forall \varphi \in \mathcal{D},$$

where $<.,.>$ is the *duality bracket* between dual topological vector spaces (such as, for example, $\mathcal{D}'$ and $\mathcal{D}$), is *commutative*, *associative* and *commutes with differentiation* (at least conditionally), so that (13) can be complemented by the following generalization of the usual *Poisson formula* for *potentials* :

(18) $$v = E * \Delta^m v.$$

A sufficient (but by no means necessary) condition for all these properties to hold is that v have a compact support in $\mathbb{R}^n$. As a matter of fact, each of the particular convolutions $\partial^\alpha E * v$ can be given a meaning provided only that v "decreases sufficiently fast at infinity", so as to match the "growth at infinity" of $\partial^\alpha E$; in view of the expression (15) of E, it follows that the sum on the right-hand side of (13) cannot serve as it is for representing all $v \in X$; surprisingly enough, this objective can be easily achieved by the simple modification of (13) we will analyze now.

An evaluation linear functional having finite support in $\mathbb{R}^n$ and annihilating $\mathcal{P} \equiv \mathcal{P}_{m-1}$ is simply a *measure* of the form

(19a) $$\mu = \sum_{k \in K} \lambda_k \delta_{(b_k)},$$

where the set K is finite, the points $b_k \in \mathbb{R}^n$ are distinct and the (real) coefficients are subject to the *orthogonality conditions*

(19b) $$< \mu, p > \equiv \sum_{k \in K} \lambda_k \, p(b_k) = 0, \ \forall p \in \mathcal{P}_{m-1};$$

by Taylor expanding the arbitrary polynomial $p \in \mathcal{P}$ about the arbitrary point $x \in \mathbb{R}^n$, it is easily seen that (19b) amounts to the following condition :

$$\text{(19b bis)} \quad \mu * p = 0, \quad \forall\, p \in \mathcal{P}_{m-1}.$$

Let us mention in passing the obviously related identity

$$\text{(20a)} \qquad (\check{v} * \varphi)(0) = \langle v,\varphi \rangle, \ \forall\, \varphi \in \mathcal{D}, \ \forall\, v \in \mathcal{D}',$$

where the symbol $\check{}$ denotes the operation of "symmetry with respect to the origin", i.e., the operation defined by

$$\text{(20b)} \qquad \check{\varphi}(x) = \varphi(-x), \ \langle \check{v},\varphi \rangle = \langle v,\check{\varphi} \rangle, \ \forall\, \varphi \in \mathcal{D}, \ \forall\, v \in \mathcal{D}';$$

this illustrates the general fact (see [14], p. 167) that duality products can be rewritten as convolution products (supposed to be meaningful !) in the so-called *trace form.*

In view of the above recalled properties of convolutions, if (13) holds valid, then we have in $\mathcal{D}'$ the following identity :

$$\text{(21)} \qquad \mu * v = \sum_{i_1,\ldots,i_m=1}^{n} \partial_{i_1\ldots i_m}(\mu * E) * \partial_{i_1\ldots i_m} v,$$

for every μ satisfying (19a,b). Now, *under assumption* (9), it turns out (see Theorem 2 below) that $\mu * E \in X$; it then follows that, for any $v \in X$, the last operation $*$ in (21) can be interpreted as a *convolution of functions* of L_2; by virtue of (a noteworthy special case of) *Young's inequality* for convolutions in Lebesgue spaces (see e.g. [17], Vol. I, pp. 115-117), the right-hand side of (21) is therefore simply a *bounded* and *uniformly continuous* function on $\mathbb{R}^n$, which *vanishes at infinity* and is such that its uniform norm is bounded by the product

$|\mu * E|_m \, |v|_m$. Moreover, it can be directly verified (by proceeding as exemplified in [14], p. 182 and p. 184), not only that each m-th partial derivative (in the distributional sense) of the right-hand side of (21) coincides in $\mathcal{D}'$ (as expected !) with the corresponding derivative of $\mu * v$, but even that (21) holds as an identity between "regular" distributions, or equivalently, as an equality almost everywhere between the associated "ordinary" functions on $\mathbb{R}^n$. We have thus finally established that, under assumption (9) and for any measure μ satisfying (19a,b), every $v \in X$ can be modified on a set of measure zero so that the *integral representation formula*

$$(22) \qquad (\mu * v)(x) = \sum_{i_1,\ldots,i_m=1}^{n} \int_{\mathbb{R}^n} [\partial_{i_1\ldots i_m}(\mu * E)](x-y) \cdot [\partial_{i_1\ldots i_m} v](y)\, dy$$

and the associated *appraisal*

$$(23a) \qquad |(\mu * v)(x)| \leqslant |\mu * E|_m \, |v|_m$$

hold everywhere on $\mathbb{R}^n$; since (23a) is essentially equivalent to Schwarz's inequality in X, it is *sharp* in the strong sense (it cannot be improved) : in actual fact, resonance occurs if, for example, $x = 0$ and v is the function $(-1)^m (\mu * E)^{\vee}$, in which case (22) reduces to

$$(23b) \qquad |\mu * E|_m^2 = (-1)^m \sum_{i\in K} \sum_{j\in K} \lambda_i \lambda_j E(b_i - b_j),$$

which is of course a significant (non-trivial !) result.

For the proof of Theorem 1 to be complete, it remains only to show that, under assumption (9) :

- the elements of X are continuous functions on $\mathbb{R}^n$ (up to possible modifications on a set of measure zero). This follows essentially from (22) and from the additional fact that *the convolution process is continuous*, at least whenever it can be regarded as a bilinear mapping of $\mathcal{D}' \times \mathcal{E}'(C)$ into $\mathcal{D}'$, where $\mathcal{E}'(C)$ denotes the space of distributions with support in a fixed compact subset C of $\mathbb{R}^n$ (see e.g. [14], p.157). For any convergent sequence $x_j \to x$ of points of $\mathbb{R}^n$, there exists of course some compact C such that $\mu_{(x_j)} \to \mu_{(x)}$ in $\mathcal{E}'(C)$, these distributions being concretely defined by (52) and satisfying (19a,b) accordingly; then $\mu_{(x_j)} * E \to \mu_{(x)} * E$ (since $E \in \mathcal{D}'$) and, by the continuity of distributional differentiation, a similar result holds in $\mathcal{D}'$ for the corresponding sequence of α-th partial derivatives (for every $\alpha \in \mathbb{N}^n$); whenever $|\alpha| = m$, convergence is known to take place in L_2, which implies by (22) and for any fixed $v \in X$ that the sequence of continuous functions $\mu_{(x_j)} * v$ on $\mathbb{R}^n$ converges pointwise (and even uniformly on compact subsets) to the continuous function $\mu_{(x)} * v$, so that finally $v(x_j) \to v(x)$ at every point.

- the seminormed space X, defined by (4) and (5), is complete, i.e., every Cauchy sequence $(v_j) \subset X$ is convergent.

As a Cauchy sequence in the Hilbert space L_2 whenever $|\alpha| = m$, $(\partial^\alpha v_j)$ converges to a uniquely defined $v^\alpha \in L_2$ while, again by (22) and for any fixed μ satisfying (19a,b), the sequence of continuous functions $\mu * v_j$ on $\mathbb{R}^n$ converges uniformly on compact subsets to a uniquely defined continuous function w. Because of the continuity of the distributional differential operators ∂^α in $\mathcal{D}'$, we must have $\partial^\alpha w = \mu * v^\alpha$ in $\mathcal{D}'$, and even in L_2 (see e.g. [14], p. 152), for all $|\alpha| = m$, so that finally $w \in X$. As a matter of fact, this argument also proves that, under the linear mapping defined by the convolution with any fixed μ satisfying (19a,b), the image of X can be interpreted as the completion of the image of $\mathcal{D}$ with respect to the seminorm $|\,.\,|_m$; on the other hand, unless $n \geqslant 2m+1$, X is a proper subspace of the completion of $\mathcal{D}$ with respect to $|\,.\,|_m$, which indeed is not even a space of distributions (see [11], p. 261).

The following result is needless to say the crux of the matter.

<u>Theorem 2</u>. *Suppose that* $m > n/2$ *and* E *is defined by* (15) *and* (16). *Then* $\mu * E \in X$ *for every measure* μ *satisfying* (19a,b).

A direct proof can be based on the representation of translations of the distribution $\ln r$ (resp. $1/r$) in the complement of the origin as *generating functions* of the Chebyshev polynomials $T_j(z)$ (resp. Legendre polynomials $P_j(z)$); we have indeed the following expansions (see e.g. [8], Vol. 2, p. 186 and p. 182) :

$$(24a)\qquad \delta_{(b)} * \ln r(x) \equiv \ln|x-b| = \ln r - \sum_{j=1}^{\infty} T_j(z)t^j/j,$$

$$(24b)\qquad \delta_{(b)} * r^{-1}(x) \equiv |x-b|^{-1} = r^{-1}\sum_{j=0}^{\infty} P_j(z)t^j,$$

where $z := \langle x,b\rangle/(r\,|b|) \in [-1,1]$ is simply the cosine of the angle θ between the elements x and b of the Euclidean space $\mathbb{R}^n$ and $t := |b|/r$ is supposed to range over a compact subinterval of $[0,1)$, both series converging then absolutely and uniformly in t and in θ (by Weierstrass'M-test, using the classical property of $|T_j(\cos\theta)|$ and $|P_j(\cos\theta)|$ to be $\leqslant 1$ for all j and θ). As readily verified, the j-th term in (24a,b) is of the form $Q_j(b)/r^j$, where $Q_j(b)$ is to be understood throughout the present proof as a *generic* notation for homogeneous polynomials of (total) degree j in the components of $b \in \mathbb{R}^n$ whose coefficients are themselves polynomials of degree $\leqslant j$ in the components of the direction vector x/r. Since $\delta_{(b)} * E(x)$ can be represented, at least for $r = |x|$ sufficiently large, as the product of the appropriate expansion (24a,b) by the $(m - [n/2])$-th power of $|x - b|^2 \equiv r^2 \{1 - 2 < x/r\ ,\ b >/r + |b|^2/r^2\}$, whose expansion for $m > n/2$ by the multinomial theorem is itself the product by that power of r^2 of a *finite* sum of terms of the form $Q_i(b)/r^i$, we finally get for any μ satisfying (19a,b) :

$$(25)\quad \mu * E(x) = \begin{cases} r^{2m-n} \ln r \sum_{i=m}^{2m-n} Q_i(b)/r^i + r^{2m-n} \sum_{j=m}^{\infty} Q_j(b)/r^j, & n \text{ even}, \\ r^{2m-n} \sum_{j=m}^{\infty} Q_j(b)/r^j, & n \text{ odd}; \end{cases}$$

as a matter of fact, it turns out here that $Q_j(b)/r^j \equiv Q_j^*(z)t^j$ is such that $Q_j^*(\cos\theta)$ is a trigonometric polynomial of degree j whose uniform norm is bounded for all j by some finite constant. We come then to the conclusion for $m > n/2$: as $r \to \infty$,

$$(26)\qquad \mu * E(x) = \begin{cases} O(r^{m-n} \ln r) + O(r^{m-n}), & \text{if } n \text{ is even}, \\ O(r^{m-n}), & \text{if } n \text{ is odd}; \end{cases}$$

hence it finally follows, by repeated application of A.A. Markov's theorem about the size of the derivatives of uniformly bounded polynomials on $[-1,1]$ (see e.g. [13], p. 105), that for $i_1, \ldots, i_m \in [1,n]$,

$$(27)\qquad \partial_{i_1 \ldots i_m} (\mu * E)(x) = \begin{cases} O(r^{-n} \ln r) + O(r^{-n}), & \text{if } n \text{ is even}, \\ O(r^{-n}), & \text{if } n \text{ is odd}, \end{cases}$$

as $r = |x| \to \infty$; the restrictions of all the partial derivatives of order m to a neighborhood of infinity, such as the complement in $\mathbb{R}^n$ of a bounded ball Ω centered at the origin and containing the support of the measure μ in its interior, are consequently square integrable. To complete the proof, it is obviously sufficient to show that the restrictions to Ω of all the m-th *distributional* partial derivatives of E are square integrable; this amounts strictly to the trivial verification that the m-th *ordinary* derivative (on the complement of the origin) of E with respect to r, that is essentially, $r^{m-n}\ln r$ if n is even (with $2m \geqslant n$) and r^{m-n} otherwise, is square integrable over Ω.

A final remark : the *finite support* assumption about μ is not strictly required for the present analysis to hold, though it is certainly suited to practical needs; what actually matters is that the P-annihilating measure μ has a *compact support*.

2.2. A Fourier Transform Approach

A classical way to look for a *fundamental solution* of the differential operator Δ^m is to make a *Fourier transform* of the *distributional* equation (14). Following L. Schwartz, the natural domain of this most powerful technique is the vector space S' of the so-called *tempered distributions*, i.e., the space of all continuous linear functionals over the space S of all infinitely differentiable functions with each derivative *rapidly decreasing at infinity*. For $\varphi \in S$, the Fourier transform $F(\varphi) \equiv \hat{\varphi}$ is defined (on $\mathbb{R}^n$) by

$$(28) \qquad \hat{\varphi}(\xi) := \int_{\mathbb{R}^n} \varphi(x)\, e^{-2i\pi \langle x,\xi \rangle}\, dx, \quad \langle x,\xi \rangle := \sum_{i=1}^{n} x_i\, \xi_i\ ;$$

it turns out that the Fourier transformation F is an *isomorphism of S onto itself*, its inverse being given by the well known *reciprocity formula :*

(29) $$\varphi(x) = \int_{\mathbb{R}^n} \hat{\varphi}(\xi) e^{2i\pi < x,\xi >} d\xi.$$

For $v \in S'$, the Fourier transform $\hat{v}$ is defined by the equation

(30) $$< \hat{v},\varphi > := < v,\hat{\varphi} > , \forall \varphi \in S.$$

As *transpose* of the linear mapping F in S, the Fourier transformation of tempered distributions is clearly an *isomorphism of S' onto itself*; moreover, it is consistent with the otherwise classical extensions of F to the Lebesgue spaces L_1 and L_2. According to the famous *Plancherel-Parseval theorem*, the Fourier transformation in the L_2 sense is a unitary isomorphism of the *complex* space L_2 onto itself, in other words, the following relations hold :

(31) $$| \hat{v} |_0 = | v |_0, \; | v |_0 := \left(\int_{\mathbb{R}^n} | v(x)|^2 dx\right)^{1/2}, \forall v \in L_2,$$

and

(32) $$(\hat{v},\hat{w})_0 = (v,w)_0, \quad \forall v,w \in L_2,$$

where $(.,.)_0$ denotes the usual Hermitian pairing in L_2 (not to be confused with the general duality bracket $<.,.>$, which is only bilinear) so that (32) can be rewritten in the form

(32 bis) $$\int_{\mathbb{R}^n} \hat{v}(\xi)\, \hat{w}(\xi)\, d\xi = \int_{\mathbb{R}^n} v(x)\, w(-x)\, dx, \forall v,w \in L_2.$$

An essential feature of the Fourier transformation of distributions is the fact that it exchanges (under certain conditions) *convolution* (of unit δ) and *multiplication* (of unit 1). This basic property is known to hold, for example, in the following cases :

- if $v \in \mathcal{E}'$ (i.e., v is a distribution with compact support in $\mathbb{R}^n$) and $w \in \mathcal{S}'$, then $v * w \in \mathcal{S}'$ and

$$(33) \qquad (v * w)\hat{} = \hat{v}.\hat{w},$$

the product of the two distributions on the right-hand side being meaningful because of the (easily proved) formula

$$(34) \qquad \hat{v}(\xi) = \langle v_x, e^{-2i\pi \langle x,\xi \rangle} \rangle, \ \forall\, v \in \mathcal{E}',$$

which shows that $\hat{v}$ is an infinitely differentiable function on $\mathbb{R}^n$. Since

$$(35) \qquad \partial^\alpha v = \partial^\alpha \delta * v, \ \forall\, v \in \mathcal{D}', \ \forall\, \alpha \in \mathbf{N}^n,$$

and again by (34),

$$(36) \qquad (\partial^\alpha \delta)\hat{}(\xi) = (2i\pi)^{|\alpha|} \xi^\alpha, \quad \forall\, \alpha \in \mathbf{N}^n,$$

with the usual definition $\xi^\alpha := \xi_1^{\alpha_1} \ldots \xi_n^{\alpha_n}$, it finally follows that

$$(37) \qquad (\partial^\alpha v)\hat{} = (2i\pi)^{|\alpha|} \xi^\alpha \, \hat{v}, \ \forall\, v \in \mathcal{S}', \ \forall\, \alpha \in \mathbf{N}^n,$$

which expresses precisely the classical fact that Fourier transformations change *derivations* into *multiplications* (by corresponding monomials) and conversely. As regards the alternative application of Fourier transformation to *partial difference equations* with constant coefficients (so important for analyzing *finite difference methods* for initial value problems), the basic relation is the following concrete application of (34) :

$$(38) \qquad (\delta_{(y)} * v)\hat{} = e^{-2i\pi \langle y,\xi \rangle} \, \hat{v}, \ \forall\, v \in \mathcal{S}',$$

it being understood again that the convolution on the left is nothing else but the transform of v by the *translation* $x \mapsto x + y$ of $\mathbb{R}^n$.

- if $u,v \in L_2$, then the Parseval identity (32bis) can be rewritten, just by making use of (38), in the apparently more general form

$$(39) \qquad \int_{\mathbb{R}^n} v(x)w(y - x)dx = \int_{\mathbb{R}^n} \hat{v}(\xi)\hat{w}(\xi)\, e^{2i\pi < y,\xi >} d\xi,$$

which indeed can be regarded as a concrete expression of (33); as already mentioned in connection with (21), the function of y defined by the convolution (39) is bounded, uniformly continuous, vanishes at infinity and its uniform norm is bounded by $|\hat{v}|_0 |\hat{w}|_0 \equiv |v|_0 |w|_0$ (by Schwarz's inequality and Plancherel's theorem).

- if $u,v \in L_1$, then (33) holds still, the Fourier transformation being here an algebra homomorphism of L_1 (with convolution) into L_∞ (with pointwise multiplication); moreover, according to the *Riemann-Lebesgue lemma*, Fourier transforms of integrable functions are continuous functions vanishing at infinity.

By virtue of *Hörmander's existence theorem in the space* $\mathcal{S}'$ (see e.g. [16], p. 314), every linear partial differential operator with constant coefficients which is not identically zero has a *tempered* fundamental solution. Solving equation (14) in $\mathcal{S}'$ (but not in $\mathcal{D}'$!) amounts therefore strictly to solving in $\mathcal{S}'$ the problem of *division of distributions*

$$(40) \qquad (-4\pi^2 \rho^2)^m \hat{E} = 1, \ \rho(\xi) := |\xi|,$$

obtained by Fourier transformation of both sides of (14). The solution of (40) is classically given by

$$(41) \qquad \hat{E} := (-4\pi^2)^{-m}\, \mathrm{Pf}\, \rho^{-2m},$$

up to the general solution $\hat{v} \in S'$ of the homogeneous equation $\rho^{2m}\,\hat{v} = 0$. As a distribution concentrated on the origin of $\mathbb{R}^n$, $\hat{v}$ is known to be simply an arbitrary linear combination of the form

$$(42) \qquad \hat{v} = \sum_{\alpha} c_{\alpha}\, \partial^{\alpha}\delta, \quad \alpha \in \mathbb{N}^n \text{ of finite length } |\alpha|;$$

moreover, it follows from Leibniz'formula that $\partial^{\alpha}\delta$ verifies the homogeneous equation iff the order relation $\alpha \geqslant 2\gamma$ (meaning that $\alpha_i \geqslant 2\gamma_i$, $1 \leqslant i \leqslant n$) does not hold for any $\gamma \in \mathbb{N}^n$ such that $|\gamma| = m$ (see [14], p. 122). As for the distribution denoted by $\mathrm{Pf}\,\rho^{\lambda}$ (where λ is a complex parameter), and often called *pseudofunction* (so as to emphasize the essential fact that its restriction to the complement of the origin in $\mathbb{R}^n$ coincides with the regular distribution defined by the locally integrable function $\xi \mapsto \rho^{\lambda}$ for $\xi \neq 0$), it can be regarded as generated by *Hadamard's finite part* "Fp" of a divergent integral according to the general rule :

$$(43) \qquad \langle \mathrm{Pf}\,\rho^{\lambda}, \varphi \rangle := \mathrm{Fp}\int_{\mathbb{R}^n} \rho^{\lambda}\varphi(\xi)d\xi, \ \forall\, \varphi \in \mathcal{D},$$

it being understood that the symbols Pf and Fp may be dropped in case of convergence (i.e., if $\mathrm{Re}\,\lambda > -n$); as readily verified (see e.g. [14],p. 122), (43) implies the important relation

$$(44) \qquad \varphi(\xi).\mathrm{Pf}\,\rho^{\lambda} = \mathrm{Pf}[\varphi(\xi).\rho^{\lambda}], \ \forall\, \varphi \in C^{\infty}(\mathbb{R}^n), \ \forall\, \lambda \text{ complex},$$

which makes obvious the fact that the distribution (41) is actually a solution of equation (40). An equivalent (though less technical) definition of the pseudofunction $\mathrm{Pf}\,\rho^{\lambda}$ is based on the essential remark that, interpreted as a regular distribution in $\mathbb{R}^n$, ρ^{λ} for $\mathrm{Re}\,\lambda > -n$ represents an *analytic* function of the parameter λ; this is equivalent with saying that, for all $\varphi \in \mathcal{D}$, the integral on the right-hand side of (43) is a *differentiable* function of the complex parameter λ varying over the domain $\operatorname{Re}\lambda > -n$. It turns out (see e.g. [9], Vol. 1, pp. 71-74 and pp. 98-99) that this regular distribution can be *analytically continued* to the entire λ-plane except for the points $\lambda = -n$, $-n-2$, $-n-4, \ldots$ at which it has first order poles, the *residue* at the pole $\lambda = -n-2k$ $(k = 0,1,\ldots)$ being the following distribution (concentrated on the origin of $\mathbb{R}^n$) :

$$(45) \qquad \operatorname{res}(\rho^{-n-2k}) = \frac{\pi^{n/2}\Delta^k\delta}{2^{2k-1}k!\,\Gamma(k+n/2)}\ , \quad k = 0,1,\ldots$$

Hence the precise definition :

- for any (complex) λ which is not of the form $-n-2k$ $(k = 0,1,\ldots)$, the pseudofunction $\operatorname{Pf}\rho^{\lambda}$ is the (unique) distribution obtained by analytic continuation of the regular distribution defined in the half-plane $\operatorname{Re}\lambda > -n$ by the locally integrable function $\xi \mapsto \rho^{\lambda}$.
- on the other hand, at each exceptional point $\lambda = -n-2k$ $(k = 0,1,\ldots)$, that is, at each pole of the meromorphic distribution-valued function of λ constructed by this analytic continuation method, $\operatorname{Pf}\rho^{\lambda}$ is the value of the regular part of its Laurent expansion about that pole.

As explained in detail in [14] (see pp. 257-258 and p. 288) and in [9] (see Vol. 1, pp. 192-195 and pp. 201-202), the inverse Fourier transformation of the pseudofunction (41) directly yields, for the corresponding (tempered) fundamental solution of Δ^m in $\mathbb{R}^n$, the expressions (15) up to the following modification : $\ln r$, in the first expression (15), is to be replaced by $(\ln r - h)$, where h is some precisely defined constant (see [14], p. 258 and p. 288); this modification is quite unimportant, however, at least in so far as we are only interested in knowing a particular fundamental solution of Δ^m, the ensuing distribution

$- \mathrm{ch}\, r^{2m-n}$ (if $2m \geq n$ and n is even) being indeed simply a *polyharmonic polynomial* (i.e., a polynomial solution of the m-th iterated Laplace equation in $\mathbb{R}^n$). As for the inverse Fourier transform of the distribution (42), it is trivially a polynomial annihilated by the operator Δ^m, so that, for our purposes, it may be dropped too. In conclusion : the general solution in S' of equation (14) is obtained by adding to the appropriate expression (15) any polyharmonic polynomial (of degree $< 2m$), while the general solution in S' of equation (40) is obtained by adding to the pseudofunction (41) any distribution (concentrated on the origin of $\mathbb{R}^n$) of the type precisely described in connection with (42).

In view of the above survey of Fourier transformation, Theorem 2, among other significant results, can be proved most simply. Let μ denote any measure satisfying (19a,b); by virtue of (34), its Fourier transform is an infinitely differentiable function, viz.,

$$(46) \qquad \hat{\mu}(\xi) = \sum_{k \in K} \lambda_k\, e^{-2i\pi < b_k, \xi >} ,$$

which is bounded on $\mathbb{R}^n$ and vanishes at the origin, together with all its partial derivatives of order $\leq m - 1$; the latter property readily follows from the fact, implied by (33) and (36), that (19b bis) amounts strictly to the conditions

$$(47) \qquad \hat{\mu}(\xi)\partial^{\alpha}\delta = 0, \ \alpha \in \mathbb{N}^n \ \text{ of length } |\alpha| \leq m - 1.$$

By the formulas (33) and (37), which indeed may be exploited here since $\mu \in E'$ and $E \in S'$, and by making use of the important property expressed by (44) in connection with the pseudodistributional part (41) of the general solution in S' of equation (40), we easily get for any $\alpha \in \mathbb{N}^n$ such that $|\alpha| = m$:

$$(48) \qquad [\partial^\alpha(\mu * E)]\hat{\ }(\xi) = (2i\pi)^{-m}\mathrm{Pf}\,[\xi^\alpha\,\hat{\mu}(\xi)\,|\xi|^{-2m}]$$

$$+ \sum_\beta c_\beta\, \xi^\alpha\, \hat{\mu}(\xi)\partial^\beta\delta,$$

where β is restricted by the condition that the order relation $\beta \geqslant 2\gamma$ may not hold in $\mathbf{N}^n$ whenever $|\gamma| = m$, so that certainly $|\beta| < 2m$. By Leibniz's formula and the key result (47), it is now readily verified that the sum on the right-hand side of (48) is identically zero; moreover, again by (47), the symbol Pf itself may be dropped in the first term, its "argument" being indeed a (bounded) integrable function on any compact neighborhood of the origin of $\mathbb{R}^n$. The ordinary function on $\mathbb{R}^n$ to which (48) thus reduces at any rate is clearly square integrable on every compact subset of $\mathbb{R}^n$; being $O(|\xi|^{-m})$ as $|\xi| \to \infty$, it is certainly in L_2 whenever $m > n/2$, which completes the proof that $\mu * E$ is then in X.

3. EXISTENCE, UNIQUENESS AND CHARACTERIZATION OF THE SOLUTION

As regards existence and uniqueness properties of the solution of Problem (P) (as stated at the end of Section 1), the following result is conclusive.

Theorem 3. *Suppose that* $m > n/2$. *Then Problem* (P) *is well-posed in the sense that its solution exists, is unique, and depends continuously on the data* $(\alpha_i)_{i\in I}$ (all other data being fixed).

The proof, together with some complementary results (important by themselves !), easily follows from a certain concrete representation formula of type (22) we will introduce presently.

In view of the $\mathcal{P}$-unisolvence of the subset $B = (a_j)_{j\in J}$ of the given finite set $A = (a_i)_{i\in I}$ of interpolation points in $\mathbb{R}^n$, there exists in $\mathcal{P} \equiv \mathcal{P}_{m-1}$ a unique basis $(p_j)_{j\in J}$ which is dual to the set of shifted Dirac measures $(\delta_{(a_j)})_{j\in J}$ (in the sense that $p_i(a_j) = \delta_{ij}$, $\forall\, i,j \in J$, where δ_{ij} is the Kronecker symbol). For every $v \in X$, and *provided that* $m > n/2$, the (uniquely defined) *$\mathcal{P}$-interpolant* Pv of v on B is accordingly given by the Lagrange formula

$$\text{(49)} \qquad Pv := \sum_{j\in J} v(a_j)p_j;$$

owing to this definition, the mapping $P : X \to X$ is a *linear projector* of X with range $\mathcal{P}_{m-1}$ and kernel

$$\text{(50)} \qquad X_0 := \{v \in X : v(a_j) = 0,\ \forall\, j \in J\},$$

so that

$$\text{(51)} \qquad X = \mathcal{P}_{m-1} \oplus X_0;$$

equipped with the seminorm $|\,.\,|_m$, X_0 is a Hilbert space (it is indeed complete, like X to which it is isometrically isomorphic, while trivially Hausdorff), the *direct sum decomposition* (51) being then topological.

By making first $x = 0$ in (22) and taking then for μ the measure-valued function on $\mathbb{R}^n$ defined by

$$\text{(52)} \qquad \check{\mu}_{(x)} := \delta_{(-x)} - \sum_{j\in J} p_j(x)\,\delta_{(-a_j)},\ \forall\, x \in \mathbb{R}^n,$$

which indeed satisfies (19a,b) everywhere, we get *for* $m > n/2$ the *direct sum representation formula* associated with (51), viz.,

$$\text{(53a)} \qquad v(x) = (Pv)(x) + (H_x, v)_m,\ \forall\, v \in X,\ \forall\, x \in \mathbb{R}^n,$$

where

(53b) $$H_x(y) := (-1)^m \left[E(x-y) - \sum_{j\in J} p_j(x)E(a_j-y) \right], \ \forall\, x,y \in \mathbb{R}^n,$$

regarded as a function of y with E defined by (15) and (16), necessarily belongs to X; by exploiting relations of type (20a,b), it is easily seen that (53a) can be rewritten in the form

(53a bis) $$< \mu_{(x)}, v > = ((-1)^m \mu_{(x)} * \check{E}, v)_m, \ \forall\, v \in X, \ \forall\, x \in \mathbb{R}^n,$$

while (53b) actually means that

(53b bis) $$H_x(y) := (-1)^m (I - P)_y\, E(x - y), \ \forall\, x,y \in \mathbb{R}^n,$$

with the standard notational convention : the subscript y appended to the linear projector I - P (where I denotes the identity mapping of X) means that this mapping is applied to E(x - y) considered as a function of y. Hence the concrete *representation formula in the Hilbert space* X_0 *for* $m > n/2$:

(54a) $$v(x) \equiv < \delta_{(x)}, v > = (K_x, v)_m, \ \forall\, v \in X_0, \ \forall\, x \in \mathbb{R}^n,$$

where K_x is the element of X_0 defined by

(54b) $$K_x := (I - P)H_x, \ \forall\, x \in \mathbb{R}^n,$$

or equivalently by

(54b bis) $$K_x(y) := (-1)^m (I - P)_x\, (I - P)_y\, E(x - y), \ \forall\, x,y \in \mathbb{R}^n.$$

By resorting again to relations of type (20a,b), we get the equivalent definition

(55a) $$K_x(y) := < (-1)^m \mu_{(x)} * \check{\mu}_{(y)}, E > , \ \forall\, x,y \in \mathbb{R}^n,$$

which is to be compared with the structurally similar equations

(55b) $E(x - y) = \langle \delta_{(x)} * \check{\delta}_{(y)}, E \rangle, \forall\, x,y \in \mathbb{R}^n,$

(55c) $H_x(y) = \langle (-1)^m \mu_{(x)} * \check{\delta}_{(y)}, E \rangle, \forall\, x,y \in \mathbb{R}^n.$

It follows from (54a) that $K_x \in X_0$ is nothing else but the (necessarily unique) *Riesz representer* of the *evaluation functional* $\delta_{(x)}$ on X_0 at the point $x \in \mathbb{R}^n$ (by Theorem 1, X_0 is indeed a *Hilbert function space* if only $m > n/2$). The set $\{K_x : \forall\, x \in \mathbb{R}^n\}$ is the so-called *reproducing kernel* of X_0, which can be regarded equivalently as the real-valued function $(x,y) \to K(x,y) \equiv K_x(y)$ on $\mathbb{R}^n \times \mathbb{R}^n$. The paramount importance of these topics, which are excellently surveyed in [3] (see pp. 316-326) and in [15] (see pp. 82-102), stems from the *reproducing property* expressed by (54a). Among other significant corollaries, the following are relevant here :

- according to the basic formula

(56) $K(x,y) = (K_x, K_y)_m, \forall\, x,y \in \mathbb{R}^n,$

the real number $K(x,y)$ can be interpreted as the inner product in X_0 of the Riesz representers of $\delta_{(x)}$ and $\delta_{(y)}$; it thus follows that $K(x,y)$ is a *symmetric* function of the two variables x,y, which is *positive definite* in the sense that, for every $N \geq 1$ and every set of points $b_1, \ldots, b_N \in \mathbb{R}^n$, the $N \times N$ matrix whose (i,j) element is $K(b_i, b_j)$ for $i,j \in [1,N]$ is positive semidefinite (it is indeed simply the *Gram matrix* of the sequence $(K_{b_i}$ $1 \leq i \leq N)$.

- the usual definition by *duality* of the so-called *negative norm* $|.|_{-m}$ (i.e., the norm in the dual space of X_0) yields the sharp appraisal

(57a) $|v(x)| \leqslant |\delta_{(x)}|_{-m} |v|_m, \ \forall\, v \in X_0, \ \forall\, x \in \mathbb{R}^n,$

where

(57b) $|\delta_{(x)}|_{-m} \equiv |K_x|_m = [K(x,x)]^{1/2}, \ \forall\, x \in \mathbb{R}^n,$

as it follows from Schwarz's inequality in X_0.

- in X_0, convergence in the mean (i.e., in the sense of the norm $|\,.\,|_m$) implies pointwise convergence and even uniform convergence on compact subsets of $\mathbb{R}^n$.

- the Riesz representer l of any continuous linear functional L on X_0 is explicitly given by

(58a) $l(x) = \langle L, K_x \rangle, \ \forall\, x \in \mathbb{R}^n,$

and its norm is such that

(58b) $|L|_{-m} \equiv |l|_m = [L_x L_y K(x,y)]^{1/2};$

similarly, if L is any bounded linear operator on X_0, then

(59) $(Lv)(x) = (L^*K_x, v)_m, \ \forall\, v \in X_0, \ \forall\, x \in \mathbb{R}^n,$

where L* denotes the *adjoint operator* of L.

Let us finally proceed to the *solution of Problem* (P). In view of the direct sum decomposition (51), the restriction of the associated projector I - P to the linear variety $V \subset X$ defined by (11) is clearly an injection. Therefore, finding $u \in V$ such that (12) holds amounts strictly to finding an element w of minimal norm $|\,.\,|_m$ in the image of V under I - P, which is the linear variety

(59a) $W := \{v \in X_0 : v(a_k) = \alpha'_k, \ \forall\, k \in K\}$

where

(59b) $K := I - J,$

(59c) $\alpha_k' := \alpha_k - \sum_{j \in J} \alpha_j \, p_j(a_k), \ \forall \, k \in K.$

As a finite intersection of translated kernels in X_0 of linearly independant bounded linear functionals $\delta_{(a_k)}$, W is a non-empty closed convex subset of the Hilbert space X_0. By virtue of the *orthogonal projection theorem*, there exists in W a unique element w of minimal norm, which is characterized by the orthogonality property :

(60a) $(w,v)_m = 0, \quad \forall \, v \in W_0,$

where

(60b) $W_0 := \{v \in X_0 : v(a_k) = 0, \ \forall \, k \in K\}.$

Since W_0 can be defined equivalently as the set of all $v \in X_0$ that are orthogonal to K_{a_k} for all $k \in K$, this optimal function w must belong to the span of $\{K_{a_k} : \forall \, k \in K\}$ and is thus necessarily of the form

(61a) $w = \sum_{i \in K} \gamma_i \, K_{a_i} \, ;$

expressing now that w belongs to W as defined by (59), we get, for determining the coefficients γ_i, the Cramer system of linear equations

(61b) $\sum_{j \in K} K(a_i, a_j) \ \gamma_j = \alpha_i', \quad \forall \, i \in K,$

whose coefficient matrix (at least if K is treated as an ordered set), is symmetric and positive definite (as Gram matrix of a sequence of linearly independent elements of X_0). The latter property implies that w, and accordingly the solution u of Problem (P) given by

$$(62) \qquad u = w + \sum_{j \in J} \alpha_j \, p_j \, ,$$

depend continuously on the data $(\alpha_i)_{i \in I}$, as claimed in Theorem 3.

A final remark of great practical significance : solving a Cramer system of linear equations with a positive definite symmetric matrix of coefficients is a most simple problem : standard algorithms, based on the idea of *Cholesky factorization*, are indeed *numerically stable* and *very economical* as regards the number of arithmetic operations; moreover, as will be explained in [12], the whole solution process of Problem (P) can be described in a *recursive* form, Newton multivariate representation formulas being accordingly given for the successive optimal interpolants.

REFERENCES

1. Ahlberg, J.H., Nilson, E.N. and Walsh, J.L. *The Theory of Splines and Their Applications*. Academic Press Inc., New York, 1967.

2. Atteia, M. Fonctions "spline" et noyaux reproduisants d'Aronszajn-Bergman, *R.I.R.O. R-3*, (1970) 31-43.

3. Davis, P.J. *Interpolation and Approximation*. Blaisdell Publishing Company, New York, 1963.

4. Deny, J. et Lions, J.-L. Les espaces du type de Beppo Levi, *Ann. Inst. Fourier V*, (1953-54) 305-370.

5. Duchon, J. Fonctions-spline à énergie invariante par rotation, *Rapport de recherche n*°27, Université de Grenoble, 1976.

6. Duchon, J. Interpolation des fonctions de deux variables suivant le principe de la flexion des plaques minces, *R.A.I.R.O. Analyse numérique 10*, (1976) 5-12.

7. Duchon, J. Splines Minimizing Rotation - Invariant Semi-Norms in Sobolev Spaces. *Constructive Theory of*

Functions of Several Variables, Oberwolfach 1976. (W. Schempp and K. Zeller ed.) 85-100. Springer-Verlag, Berlin, Heidelberg, 1977.

8. Erdélyi, A., Magnus, W., Oberhettinger, F. and Tricomi, F.G. *Higher Transcendental Functions,* 3 volumes. McGraw-Hill Book Company, Inc., New York, 1953.

9. Gel'fand, I.M. and Shilov, G.E. *Generalized Functions,* 5 volumes. Academic Press Inc., New York, 1964.

10. Harder, R.L. and Desmarais, R.N. Interpolation Using Surface Splines, *J. Aircraft 9*, (1972) 189-191.

11. Hörmander, L. et Lions, J.-L. Sur la complétion par rapport à une intégrale de Dirichlet, *Math. Scand. 4*, (1956) 259-270.

12. Meinguet, J. Multivariate Interpolation at Arbitrary Points Made Simple, *ZAMP*, to appear. (1979)

13. Rivlin, T.J. *The Chebyshev Polynomials.* John Wiley & Sons, Inc., New York, 1974.

14. Schwartz, L. *Théorie des Distributions.* Hermann, Paris, 1966.

15. Shapiro, H.S. *Topics in Approximation Theory.* Springer-Verlag, Berlin, Heidelberg, 1971.

16. Trèves, F. *Linear Partial Differential Equations with Constant Coefficients.* Gordon and Breach Science Publishers,Inc., New York, 1966.

17. Vo-Khac Khoan. *Distributions, Analyse de Fourier, Opérateurs aux Dérivées Partielles,* 2 volumes, Vuibert, Paris, 1972.

INEQUALITIES OF MARKOFF AND BERNSTEIN

Q.I. Rahman
Department of Mathematics
University of Montreal
Montreal, Canada H3C 3J7

Abstract: In the early part of this century, de la Vallée Poussin raised the following question of best approximation: Is it possible to approximate every polygonal line by polynomials of degree n with an error of $o(1/n)$ as n becomes large? The question was answered in the negative by S. Bernstein. For this he proved and made considerable use of an inequality concerning the derivatives of polynomials. This inequality and a related (and earlier) one by A.A. Markoff have been the starting point of a considerable literature. Here we discuss some of the investigations which have centered about these inequalities.

In the year 1889 A.A. Markoff [12] proved the following

THEOREM 1. *If* $P_n(x) = \sum_{\nu=0}^{n} a_\nu x^\nu$ *is a polynomial of degree* n *such that* $|P_n(x)| \leq 1$ *on the interval* $-1 \leq x \leq 1$, *then*

$$\max_{-1\leq x\leq 1} |P_n'(x)| \leq n^2. \tag{1}$$

The n-th Tchebycheff polynomial of the first kind

$$T_n(x) = \cos(n \text{ arc} \cos x) = 2^{n-1} \prod_{\nu=1}^{n} \left\{x - \cos((\nu - \tfrac{1}{2})\pi/n)\right\} \tag{2}$$

satisfies the conditions of Theorem 1 and its derivative at the point $x = 1$ is equal to n^2. So the constant n^2 in (1) cannot be replaced by any lower constant.

The case $n = 2$ of this theorem was already known to the

Badri N. Sahney (ed.), Polynomial and Spline Approximation, 191-201.

chemist Mendeleieff (author of the periodic table of elements) who needed it in connection with his work in chemistry.

W.A. Markoff (brother of A.A. Markoff) considered the problem of determining exact bounds for the k-th derivative of $P_n(x)$ at a given point x_0 in [-1,1] under the conditions of Theorem 1. His results appeared in a Russian journal in the year 1892; a German version of his remarkable paper was later published in Mathematische Annalen [13]. Amongst other things he proved:

THEOREM 2. *Under the conditions of Theorem* 1

$$(3) \quad \max_{-1\le x\le 1} |P_n^{(k)}(x)| \le \frac{n^2(n^2-1^2)(n^2-2^2)\dots(n^2-(k-1)^2)}{1\cdot 3\cdot 5\dots(2k-1)}, \qquad k = 1,2,\dots,n.$$

The right-hand side of this inequality is exactly equal to $T_n^{(k)}(1)$, where $T_n(x)$ is the n-th Tchebycheff polynomial of the first kind (2).

It was shown by Duffin and Schaeffer [6] that for (3) it is enough to assume that $|P_n(x)| \le 1$ at the $n+1$ points $x = \cos(k\pi/n)$; $k = 0,1,2,\dots,n$.

There is an analogue of (1) for trigonometric polynomials which is known as Bernstein's inequality. It may be stated as follows.

THEOREM 3. *If* $t(\theta) = \sum_{\nu=-n}^{n} a_\nu e^{i\nu\theta}$ *is a trigonometric polynomial of degree n and* $|t(\theta)| \le 1$ *for real* θ, *then*

$$(4) \quad |t'(\theta)| \le n, \qquad \theta \text{ real}.$$

In (4) equality holds if and only if $t(\theta) = e^{i\gamma}\cos(n\theta-\alpha)$ where γ and α are arbitrary real numbers.

If $P_n(x) = \sum_{\nu=0}^{n} a_\nu x^\nu$ is a polynomial of degree n such that $|P_n(x)| \le 1$ on [-1,1] then $P_n(\cos\theta)$ is a trigonometric polynomial $t(\theta)$ of degree n such that $|t(\theta)| \le 1$ for all real θ and so Bernstein's inequality (4) implies that

$$(5) \quad |P_n'(x)| \le \frac{n}{\sqrt{1-x^2}} \quad \text{for } -1 < x < 1.$$

This inequality which is also known as Bernstein's inequality first appeared in a prize-winning essay [1] on problems of best approximation. It gives a much better estimate of $|P_n'(x)|$ than does A.A. Markoff's inequality (1) except for a small neighbourhood of the points -1 and +1. In fact, it is not hard to prove (1) if (5) is already known.

Although (1) is sharp it follows from (5) that for a fixed x in the open interval (-1,1), $|P_n'(x)| = O(n)$ as n goes to infinity. The underlying reason was discovered by Szegö [21]. Let Γ represent an open or a closed Jordan curve in the complex z-plane, and at any point z_0 on Γ let $\alpha\pi$ be the exterior angle. In the case in which Γ is an open curve there will of course be two exterior angles at each point z_0 except the ends. In this case let $\alpha\pi$ be the larger of the two exterior angles at z_0. With suitable smoothness conditions on Γ we have the following:

THEOREM 4. *If* $P_n(z) = \sum_{\nu=0}^{n} a_\nu z^\nu$ *is a polynomial of degree* n *such that* $|P_n(z)| \le 1$ *on* Γ, *then*

$$(6) \quad |P_n'(z_0)| \le cn^\alpha, \quad z_0 \in \Gamma.$$

Here c *is a constant which depends on* z_0 *and* Γ *but not on* n.

The bound cn^α in this inequality is of the precise order as n becomes infinite. In the case in which Γ is the unit interval [-1,1] we see that $\alpha = 2$ at the endpoints of the interval, so this theorem shows that the derivative is of order n^2. On the other hand at points in the interior of the interval $\alpha = 1$, so at these points the derivative is of order n.

A trigonometric polynomial $t(z) = \sum_{\nu=-n}^{n} a_\nu e^{i\nu z}$ is an entire function of the complex variable z. In fact, it is an entire function of exponential type n. Let us recall that an entire function $f(z)$ is said to be of exponential type τ if for every $\varepsilon > 0$,

$$|f(z)| = O\left(e^{(\tau+\varepsilon)|z|}\right)$$

for all z in the complex plane. This means that entire functions of exponential type τ include all functions of order 1 and type less than or equal to τ, as well as all functions of order less than 1. Bernstein [2] found that the condition in Theorem 3 that $f(z)$ is a trigonometric polynomial can be replaced by the milder condition that it is an entire function of exponential type. In fact, the following theorem holds:

THEOREM 5. *If $f(z)$ is an entire function of exponential type τ and $|f(x)| \le 1$ for all real x, then*

(7) $|f'(x)| \le \tau.$

Here equality holds if and only if $f(z) = e^{i\gamma}\cos(\tau z-\alpha)$.

The inequalities in the formulation of Theorem 5 may be written in the form

$$|f(x)| \le |e^{i\tau x}| \text{ and } |f'(x)| \le |(e^{i\tau x})'|,$$

i.e. an inequality between certain entire functions of exponential type is preserved under the operation of differentiation. This consideration led to a very significant generalization of Theorem 5. But before we state it we need to introduce some definitions.

DEFINITION 1. An entire function $\omega(z)$ of exponential type having no zeros in the lower half-plane and satisfying $|\omega(x+iy)| \ge |\omega(x-iy)|$ for $y < 0$ is said to belong to the class P.

DEFINITION 2. An additive homogeneous operator $B[f(z)]$ which carries entire functions of exponential type into entire functions of exponential type and leaves the class P invariant is called a B-operator.

Differentiation is a B-operator and so Theorem 5 is a special case of the following theorem due to Levin [10].

THEOREM 6. *If $f(z)$ is an entire function of exponential type τ, B is a B-operator, and $\omega(z)$ is an entire function of class P and of order 1, type $\sigma \ge \tau$, then*

(8) $|f(x)| \le |\omega(x)|, \quad -\infty < x < \infty,$

implies

(9) $|B[f(x)]| \le |B[\omega(x)]|, \quad -\infty < x < \infty.$

The relevance of class P is shown by the fact that (8) does not imply (9) for all $f(z)$ of exponential type unless $\omega(z)$ or $\overline{\omega}(z)$ belongs to P.

THEOREM 7. *Let $\omega(z)$ be an entire function of order 1 and* *type* σ *such that neither* $\omega(z)$ *nor* $\overline{\omega}(z)$ *belongs to P. Then there is an entire function* $f(z)$ *of exponential type* $\tau \le \sigma$ *such that* $|f(x)| \le |\omega(x)|$ *for* $-\infty < x < \infty$ *while for some real* x_0, $|f'(x_0)| > |\omega'(x_0)|$.

Here is an interesting special case of Theorem 3 which deserves to be mentioned separately.

THEOREM 8. *If* $P_n(z) = \sum_{\nu=0}^{n} a_\nu z^\nu$ *is a polynomial of degree n such that* $|P_n(z)| \le 1$ *on* $|z| = 1$ *then*

$$(10) \quad |P_n'(z)| \le n \quad \text{for } |z| \le 1.$$

Here equality holds only for $P_n(z) = e^{i\gamma}z^n$, i.e. *when all the zeros of* $P_n(z)$ *lie at the origin.*

It was conjectured by P. Erdös and proved by Lax [9] that if $P_n(z) \ne 0$ in $|z| < 1$, then (10) can be replaced by

$$(11) \quad \max_{|z|\le 1} |P_n'(z)| \le n/2$$

where equality holds if all the zeros of $P_n(z)$ lie on $|z| = 1$. For alternative proofs see [5]. It was extended to entire functions of exponential type by Boas [4].

In the other direction, there is a theorem of Turán [22] which states that if $P_n(z)$ is a polynomial of degree n having all its zeros in $|z| \le 1$, then

$$(12) \quad \max_{|z|\le 1} |P_n'(z)| \ge (n/2) \max_{|z|\le 1} |P_n(z)|.$$

Several years ago Prof. R.P. Boas asked: *how large can*

$$(13) \quad \left\{\max_{|z|\le 1} |P_n'(z)|\right\} \Big/ \left\{\max_{|z|\le 1} |P_n(z)|\right\}$$

be if $P_n(z)$ *is an arbitrary polynomial of degree n not vanishing in* $|z| < k$ *where k is a given positive number?* A partial answer was given by Malik [11] who proved the following

THEOREM 9. *If* $P_n(z)$ *is a polynomial of degree n such that*

$|P_n(z)| \le 1$ *for* $|z| = 1$, *and* $P_n(z) \ne 0$ *in* $|z| < k$, *then*

(14) $|P_n'(z)| \le n/(1+k)$ *for* $|z| \le 1$,

provided $k \ge 1$.

For $k = 1$ it reduces to (11). If $k > 1$, then in (14) equality holds if and only if $p_n(z)$ is of the form $e^{i\gamma}\left(\frac{z+ke^{i\alpha}}{1+k}\right)^n$.

In the case $k < 1$, the precise answer to the above mentioned question of Boas is not known.

Here is another question which was asked by Boas: *how large can* (13) *be if* $P_n(z)$ *is an arbitrary polynomial of degree n having exactly k zeros on or outside the unit circle*?

This latter problem appears to be very difficult and the precise answer is not known even in the special case $k = 1$. There is something rather surprising in this connection: In view of (10),(11), one might suspect that if $P_n(z)$ is known to have a zero on or outside the unit circle then $\max_{|z|\le 1} |P_n'(z)| \le (n-c) \max_{|z|\le 1} |P_n(z)|$ where c is a constant possibly equal to $\frac{1}{2}$. But that is far from the truth as the following theorem [7] shows.

THEOREM 10. *For every positive integer n there exists a polynomial* $P_n(z)$ *of degree n having a zero on* $|z| = 1$ *such that*

$$(15) \quad \max_{|z|\le 1} |P_n'(z)| \ge \left(n - \frac{c_1}{n}\right) \max_{|z|\le 1} |P_n(z)|$$

where c_1 *is a constant independent of n. On the other hand, for an arbitrary polynomial* $P_n(z)$ *of degree n having a zero on* $|z| = 1$ *we have*

$$(16) \quad \max_{|z|\le 1} |P_n'(z)| \le \left(n - \frac{1-\sin 1}{4\pi}\frac{1}{n}\right) \max_{|z|\le 1} |P_n(z)|.$$

L^p inequalities. The following generalization of Theorem 3 was obtained by A. Zygmund [23].

THEOREM 11. *Let* $p \ge 1$. *If* $t(\theta)$ *is a trigonometric polynomial of degree n, then*

(17) $$\left(\frac{1}{2\pi}\int_{-\pi}^{\pi}|t'(\theta)|^p d\theta\right)^{1/p} \le n\left(\frac{1}{2\pi}\int_{-\pi}^{\pi}|t(\theta)|^p d\theta\right)^{1/p}.$$

Theorem 3 can be obtained from this by letting p tend to infinity. A similar generalization of (11) was obtained by de Bruijn who proved [5]:

THEOREM 12. *If $P_n(z)$ is a polynomial of degree n such that $P_n(z) \neq 0$ in $|z| < 1$, then*

(18) $$\left(\frac{1}{2\pi}\int_{-\pi}^{\pi}|P_n'(e^{i\theta})|^p d\theta\right)^{1/p} \le n\, C_p\left(\frac{1}{2\pi}\int_{-\pi}^{\pi}|P_n(e^{i\theta})|^p d\theta\right)^{1/p}, \ p \ge 1,$$

where

$$C_p = \left(\frac{1}{2\pi}\int_{-\pi}^{\pi}|1 + e^{i\phi}|^p d\phi\right)^{-1/p}.$$

Corresponding to Theorem 10, we have the following result [7] in the special case $p = 2$.

THEOREM 13. *If $P_n(z)$ is a polynomial of degree n having a zero on $|z| = 1$, then*

(19) $$\frac{1}{2\pi}\int_{-\pi}^{\pi}|P_n'(e^{i\theta})|^2 d\theta \le (n^2 - \alpha_n)\frac{1}{2\pi}\int_{-\pi}^{\pi}|P_n(e^{i\theta})|^2 d\theta$$

where α_n is the unique root of the equation

(20) $$\sum_{\nu=0}^{n}\frac{1}{n^2-\nu^2-x} = 0$$

in the interval $(0, 2n-1)$.

It can be shown that $\alpha_n \sim 2n/(\log n)$ as $n \to \infty$.

There is an L^p analogue of Theorem 5 too. It states that if $f(z)$ is an entire function of exponential type τ belonging to L^p on the real axis, then [3, p.211]

(21) $$\int_{-\infty}^{\infty}|f'(x)|^p dx \le \tau^p\int_{-\infty}^{\infty}|f(x)|^p dx, \ p \ge 1.$$

Note that (7) does not follow from (21). For a similar result corresponding to Theorem 12 see [19].

Some further open problems. In addition to some of the unanswered questions already mentioned, here are a few others.

1. Let $\varphi(x) \geq 0$ for $-1 \leq x \leq 1$ and consider the class $P_{n,\varphi}$ of all polynomials $P_n(x) = \sum_{\nu=0}^{n} a_\nu x^\nu$ of degree at most n such that $|P_n(x)| \leq \varphi(x)$ for $-1 \leq x \leq 1$. How large can $|P_n^{(k)}(x_0)|$ be at a given point $x_0 \in [-1,1]$ if $P_n(x)$ is an arbitrary polynomial belonging to $P_{n,\varphi}$?

This problem was raised by P. Turán at a conference held in Varna, Bulgaria in the year 1970. He singled out the case $\varphi(x) = \sqrt{1-x^2}$ as one of special interest since (for polynomials which are real for real x) the condition means that the graph of $P_n(x)$ on $[-1,1]$ lies in the closed unit disk. The case $\varphi(x) \equiv 1$ is of course classical and was thoroughly investigated by W. Markoff. In the case $\varphi(x) = \sqrt{1-x^2}$ it is known ([18], [15], [16]) that for $n \geq 2$ and $k = 1,2,3,\ldots$

$$(22) \qquad \max_{-1\leq x\leq 1} |P_n^{(k)}(x)| \leq \left|\frac{d^k}{dx^k}\left\{(1-x^2)U_{n-2}(x)\right\}\right|_{x = \pm 1}$$

where $U_{n-2}(x)$ is the Tchebycheff polynomial of the second kind of degree $n - 2$.

Nothing of interest is known in the general case. As regards (22) it may be asked if it is enough to assume that $|P_n(x)| \leq \sqrt{1-x^2}$ at -1, $+1$ and at the $n - 1$ points $\cos\frac{\pi(2k+1)}{2(n-1)}$, $k = 0,1,2,\ldots,n-2$.

2. It has recently been proved by Newman [14] that if Λ is a finite set of positive numbers λ then for an arbitrary expression $\pi(x) = c_0 + \sum_{\Lambda} c_\lambda x^\lambda$ we have

$$(23) \qquad \max_{[0,1]} |x\pi'(x)| \leq 11(\sum_{\Lambda}\lambda) \max_{[0,1]} |\pi(x)|.$$

In particular

$$\max_{[\frac{1}{2},1]} |\pi'(x)| \le 22(\sum_{\Lambda}\lambda) \max_{[0,1]} |\pi(x)|.$$

Hence, if Λ consists of the integers $1,2,\ldots,n$, then

$$\max_{[\frac{1}{2},1]} |\pi'(x)| \le 11n(n+1) \max_{[0,1]} |\pi(x)|.$$

Equivalently, if $P_n(x)$ is a polynomial of degree n such that $\max_{[-1,1]} |P_n(x)| \le 1$, then for $0 \le x \le 1$

$$(24) \quad |P_n'(x)| \le \frac{11}{2}\, n(n+1) \sim \frac{11}{2}\, n^2.$$

By symmetry, (24) must hold for $-1 \le x \le 0$ as well. Note that (24) differs from (1) only by a constant factor and thus (23) consitutes a nice and useful extension of Theorem 1 except that the constant 11 on the right is not the best possible. It will be of interest to get the best constant.

3. It was proved by Pommerenke [17] that if $P_n(z)$ is a polynomial of degree n such that $|P_n(z)| \le 1$ on a connected, closed, bounded set E of capacity κ, then for all $z \in E$

$$|P_n'(z)| \le \frac{e}{2}\,\frac{n^2}{\kappa}.$$

Here the constant $\frac{e}{2}$ cannot be replaced by any constant less than $\frac{1}{2}$. But what is the best possible constant?

4. It was proved by Sewell [20] that if E is an ellipse with semi-axes a and b, $a \ge b$, then for every polynomial $P_n(z)$ of degree n

$$\max_{z\in E} |P_n'(z)| \le \frac{n}{b} \max_{z\in E} |P_n(z)|.$$

This is not quite satisfactory and there must be a better inequality than this.

5. The following problem was mentioned to me by Professor E.B. Saff.

Let $P_n(z) = \prod_{\nu=1}^{n} (z-z_\nu)$ be a polynomial having all its zeros in $Re z \ge 1$. Is it true that

$$\max_{|z|=1} |P'_n(z)| \le \sum_{\nu=1}^{n} \frac{1}{1+Re z_\nu} \max_{|z|=1} |P_n(z)|?$$

This is known to be true [8] for $n = 1, 2$ and for arbitrary n if $P_n(z)$ is real for real z.

6. How large can (13) be if $P_n(z)$ is an arbitrary polynomial of degree n satisfying $z^n P_n(1/z) \equiv P_n(z)$?

REFERENCES

1. S.N. Bernstein, *Sur l'ordre de la meilleure approximation des fonctions continues par des polynomes de degré donné*, Mémoire de l'Académie Royale de Belgique, (2), vol. 4 (1912), pp. 1-103.

2. S.N. Bernstein, *Sur une propriété des fonctions entières*, Comptes Rendus de l'Académie des Sciences, Paris, vol. 176 (1923), pp. 1603-1605.

3. R.P. Boas, Jr., *Entire functions*, Academic Press, New York, 1954.

4. R.P. Boas, Jr., *Inequalities for asymmetric entire functions*, Illinois J. Math., vol. 1 (1957), pp. 94-97.

5. N.G. de Bruijn, *Inequalities concerning polynomials in the complex domain*, Nederl. Akad. Wetensch., Proc. vol. 50 (1947), pp. 1265-1272 = Indag. Math., vol. 9 (1947), pp. 591-598.

6. R.J. Duffin and A.C. Schaeffer, *A refinement of an inequality of the brothers Markoff*, Trans. Amer. Math. Soc., vol. 50 (1941), pp. 517-528.

7. A. Giroux and Q.I. Rahman, *Inequalities for polynomials with a prescribed zero*, Trans. Amer. Math. Soc., vol. 193 (1974), pp. 67-98.

8. A. Giroux, Q.I. Rahman and G. Schmeisser, *On Bernstein's inequality*, Canadian J. Math. (to appear).

9. P.D. Lax, *Proof of a conjecture of P. Erdös on the derivative of a polynomial*, Bull. Amer. Math. Soc., vol. 50 (1944), pp. 509-513.

10. B.Ya. Levin, *On a special class of entire functions and on related extremal properties of entire functions of finite degree*, Izvestiya Akad. Nauk SSSR. Ser. Mat., vol. 14(1950), pp. 45-84.

11. M.A. Malik, *On the derivative of a polynomial*, J. London Math. Soc., vol. 1 (1969), pp. 57-60.

12. A.A. Markoff, *Sur une question posée par Mendeleieff*, Bulletin of the Academy of Sciences of St. Petersburg, vol. 62 (1889), pp. 1-24.

13. W.A. Markoff, *Über Polynome die in einen gegebenen Intervalle möglichst wenig von Null abweichen*, Math. Annalen, vol. 77 (1916), pp. 213-258.

14. D.J. Newman, *Derivative bounds for Müntz polynomials* , J. Approx. Theory, vol. 18 (1976), pp. 360-362.

15. R. Pierre and Q.I. Rahman, *On a problem of Turán about polynomials*, Proc. Amer. Math. Soc., vol. 56 (1976), pp. 231-238.

16. R. Pierre, *Sur les polynômes dont le graphe sur* [-1,+1] *se trouve dans une région convexe donnée*, Ph.D. thesis, Université de Montréal, 1977.

17. C. Pommerenke, *On the derivative of a polynomial*, Mich. Math. J., vol. 6 (1969), pp. 373-375.

18. Q.I. Rahman, *On a problem of Turán about polynomials with curved majorants*, Trans. Amer. Math. Soc., vol. 163 (1972), pp. 447-455.

19. Q.I. Rahman, *On asymmetric entire functions. II*, Math. Annalen, vol. 167 (1966), pp. 49-52.

20. W.E. Sewell, *On the polynomial derivative constant for an ellipse*, Amer. Math. Monthly, vol. 44 (1937), pp. 577-578.

21. G. Szegö, *Über einen Satz von A. Markoff*, Math. Z., vol. 23 (1925), pp. 46-61.

22. P. Turán, *Über die Ableitung von Polynomen*, Compositio Math., vol. 7 (1939-40), pp. 89-95.

23. A. Zygmund, *A remark on conjugate series*, Proc. London Math. Soc., (2), vol. 34 (1932), pp. 392-400.

SIMULTANEOUS INTERPOLATION AND APPROXIMATION

R. Gervais, Q.I. Rahman* and G. Schmeisser

Abstract: Given the values of a function and possibly the values of some of its derivatives, at certain points, a practical problem of numerical analysis is to use this information to construct other functions which approximate it. Simultaneous interpolation and approximation of continuous functions on a compact interval, by polynomials, has been extensively studied by Runge, Bernstein, Faber, Fejér, Turán and others. Here we study simultaneous interpolation and approximation of a function f on the whole real line by entire functions of exponential type. The function f is supposed to be uniformly continuous and bounded on $(-\infty,\infty)$.

1. Introduction

1.1. Given a continuous function f

$$[-1,1] \xrightarrow{f} R$$

it is natural to think that the sequence of polynomials $\{p_n(f;\cdot)\}$ which duplicate the function at $n+1$ equally spaced points of the interval will converge uniformly to f as $n \to \infty$. But this may not be the case as was first noted by Runge [15]. Then Bernstein [2] observed that the sequence of polynomials $\{p_n(|x|;\cdot)\}$ interpolating the function $|x|$ in the $n+1$ equally spaced points

$$-1 + \frac{2\nu}{n} \qquad (\nu = 0,1,2,\ldots,n)$$

converges to $|x|$ at no point of $[-1,1]$ except the points -1, 0, and 1. The possibility that a more appropriate choice of interpolation points would give rise to a convergent interpolation process was ruled out when Faber [8] proved

THEOREM A. *Given an infinite triangular matrix*

*Being invited speaker, Professor Rahman presented this paper.

Badri N. Sahney (ed.), Polynomial and Spline Approximation, 203–223.

$$A = \begin{pmatrix} x_{1,1} & & & \\ x_{2,1} & x_{2,2} & & \\ \vdots & & & \\ x_{n,1} & x_{n,2} & \cdots & x_{n,n} \\ \vdots & \vdots & & \end{pmatrix}$$

where

$$(1.1) \qquad 1 \geq x_{n,1} > x_{n,2} > \cdots > x_{n,n} \geq -1$$

for $n = 1,2,\ldots$, *there exists a continuous function* f *such that the associated sequence of Lagrange interpolating polynomials*

$$L_n(f;x) = \omega_n(x) \sum_{\nu=1}^{n} \frac{1}{\omega_n'(x_{n,\nu})} \frac{f(x_{n,\nu})}{x-x_{n,\nu}}, \quad \left(\omega_n(x) = \prod_{\nu=1}^{n} (x-x_{n,\nu})\right)$$

does not converge uniformly to f *on* $[-1,1]$.

Proof. There exists a polynomial $p_{n-1}(x) = \sum_{\nu=0}^{n-1} a_\nu x^\nu$ such that ([8], for another proof see [9, pp. 450–453])

$$|p_{n-1}(x_{n,\nu})| \leq 1 \qquad (\nu = 1,2,\ldots,n)$$

whereas

$$|p_{n-1}(x_{n,n+1})| > \frac{1}{12} \log (n-1)$$

at some point $x_{n,n+1} \in [-1,1]$. We can clearly construct a continuous function f_n on $[-1,1]$ such that

(i) $f_n(x_{n,\nu}) = p_{n-1}(x_{n,\nu}) \qquad (\nu = 1,2,\ldots,n)$

(ii) $|f_n(x)| \leq 1$ for $-1 \leq x \leq 1$.

Thus

$$L_n(f_n;x) = L_n(p_{n-1};x) = p_{n-1}(x) .$$

Now note that

$$f \to L_n(f;\cdot)$$

defines a bounded linear transformation L_n from the Banach space $C[-1,1]$ of all continuous functions f

$$[-1,1] \xrightarrow{f} R$$

with $\|f\| = \max\limits_{-1\le x\le 1} |f(x)|$ into itself. From above

$$\|L_n\| \ge \|L_n(f_n;\cdot)\| = \|p_{n-1}\| \ge |p_{n-1}(x_{n,n+1})| > \frac{1}{12}\log(n-1) \quad ,$$

i.e. $\sup\limits_n \|L_n\| = \infty$. Hence the theorem of Faber follows from

THE BANACH-STEINHAUS THEOREM [14, p. 98]. *Suppose X is a Banach space, Y is a normed linear space, and $\{\Lambda_\alpha\}$ is a collection of bounded linear transformations of X into Y, where α ranges over some index set I. Then either there exists an $M < \infty$ such that*

$$\|\Lambda_\alpha\| \le M$$

for every $\alpha \in I$, or

$$\sup_{\alpha\in I} \|\Lambda_\alpha \xi\| = \infty$$

for some $\xi \in X$.

Bernstein [3] even proved that given an arbitrary A with (1.1) there exists a function $f_0 \in C[-1,1]$ and a point $x_0 \in [-1,1]$ such that

$$\limsup_{n\to\infty} |L_n(f_0;x_0)| = \infty \quad .$$

It was discovered by Fejér that the situation changes drastically if instead of Lagrange interpolation we consider an appropriate special case of the general Hermite interpolation. He proved [9] that if A is the Tchebycheff matrix, i.e. the points $x_{n,\nu}$ are taken to be the zeros of the Tchebycheff polynomial $T_n(x) = \cos n\,(\text{arc}\cos x)$,

$$(1.2) \qquad x_{n,\nu} = \cos\frac{2\nu-1}{2n}\pi \qquad (\nu = 1,2,\ldots,n) \quad ,$$

then for every continuous function f the sequence $\{H_n(f;\cdot)\}$ of polynomials of degree $2n-1$ satisfying the conditions

$$\left.\begin{aligned} H_n(f;x_{n,\nu}) &= f(x_{n,\nu}) \\ H_n'(f;x_{n,\nu}) &= y_{n,\nu}' \end{aligned}\right\} \quad (\nu = 1,2,\ldots,n)$$

converges uniformly to f on $[-1,1]$ provided

$$\lim_{n\to\infty} \max_{1\le\nu\le n} \frac{|y_{n,\nu}'| \log n}{n} = 0 . \tag{1.3}$$

It was shown by Marcinkiewicz [13] that for the Lagrange interpolation process the Tchebycheff matrix can be just as bad as any other. In fact, he proved that there exists a continuous function f on $[-1,1]$ for which the Lagrange interpolation process in the nodes (1.2) diverges at all points of the interval $[-1,1]$.

Turán and others investigated the behaviour of (0,2)-interpolating polynomials, i.e. polynomials $R_n(f;\cdot)$ of degree $2n-1$ which duplicate the function f at the n points (1.1) and whose second derivative assumes prescribed values $y_{n,\nu}''$ at these points. Contrary to $H_n(f;\cdot)$ the polynomials $R_n(f;\cdot)$ do not necessarily exist and may not be unique if they exist. If the points $x_{n,\nu}$ in (1.1) are taken to be the zeros of the polynomial

$$\pi_n(x) = (1-x^2)P_{n-1}'(x)$$

where P_{n-1} is the Legendre polynomial of degree $n-1$, then there exists a unique (0,2)-interpolating polynomial $R_n(f;\cdot)$ provided n is even, say $n = 2m$. Further, if f is continuously differentiable in $[-1,1]$ with the modulus of continuity $\omega(\delta,f')$ of f' such that

$$\int_0 \frac{\omega(t,f')}{t}\,dt$$

exists and the numbers $y_{2m,\nu}''$ satisfy

$$\max_{1\le\nu\le 2m} |y_{2m,\nu}''| = o(m) \quad ,$$

then [1] the sequence $\{R_{2m}(f;\cdot)$ converges uniformly to f on $[-1,1]$ as m goes to ∞. Kiš [12] examined the analogous problem for (0,2)-interpolation in the roots of unity.

1.2. If f is continuous and bounded on the whole real axis then a polynomial cannot interpolate it in an infinite set of points x_ν, $\nu = 0,\pm 1,\pm 2,\ldots$ such that $\lim_{\nu\to\pm\infty} x_\nu = \pm\infty$. Besides, rational functions can be used to approximate only those continuous functions arbitrarily closely whose graphs happen to have asymptotes parallel to the real axis. But we can use entire functions.

It was proved by Carleman [6] that a continuous function

$$(-\infty,\infty) \xrightarrow{f} \mathbb{C}$$

can be approximated arbitrarily closely on $(-\infty,\infty)$ by entire functions; in fact,

$$\sup |f(x)-g(x)|\ \varphi(|x|)$$

can be made arbitrarily small, with g entire, no matter how fast the given (continuous and nonnegative) function φ grows. However, the approximating entire function is generally of infinite order.

Let us recall that an entire function g is said to be of order ρ if

$$\limsup_{r\to\infty} \frac{\log\log M(r)}{\log r} = \rho \qquad (0 \le \rho \le \infty)$$

where $M(r) = \max_{|z|=r} |g(z)|$. The entire function g of positive order ρ is of type τ if

$$\limsup_{r\to\infty} \frac{\log M(r)}{r^\rho} = \tau \qquad (0 \le \tau \le \infty)\ .$$

A function g is said to be of growth (ρ,τ) if it is of order not exceeding ρ and of type not exceeding τ if of order ρ. A function of growth $(1,\tau)$, $\tau < \infty$, is called a function of exponential type τ. Thus g is an entire function of exponential type τ if for every $\varepsilon > 0$

$$|g(z)| = O(e^{(\tau+\varepsilon)|z|}) \tag{1.4}$$

for all $z \in \mathbb{C}$. If (1.4) holds in an unbounded connected subset E of the complex plane, then g is said to be of exponential type

τ in E.

According to Bernstein's inequality if g is an entire function of exponential type τ such that

$$|g(x)| \leq M \quad \text{on } (-\infty,\infty) , \tag{1.5}$$

then

$$|g'(x)| \leq M\tau \quad \text{on } (-\infty,\infty) , \tag{1.6}$$

i.e. (1.5) implies uniform continuity on the whole real axis.

It was shown by Bernstein [4] that if f is continuous and bounded on the whole real axis and $A_\tau[f]$ denotes the minimum of $\sup_{-\infty<x<\infty} |f(x)-g_\tau(x)|$ for all entire functions g_τ of exponential type τ, then $A_\tau[f] \to 0$ as $\tau \to \infty$ if and only if f is uniformly continuous on $(-\infty,\infty)$. Let us agree to denote by $UCB(-\infty,\infty)$ the class of all functions which are uniformly continuous and bounded on $(-\infty,\infty)$.

Problems of simultaneous interpolation and approximation by entire functions of exponential type, analogous to those described in §1.1 have received little attention. There are various questions which need to be answered first:

(i) Given an infinite set of interpolation points on the real line does there always exist an entire function *of exponential type* which interpolates a given function in these points? This is not always true and the answer depends on the given function and the distribution of the interpolation points.

(ii) If for a given set of interpolation points $\{\lambda_n\}_{n=-\infty}^{\infty}$ there do exist entire functions of exponential type which duplicate an arbitrary function $f \in UCB(-\infty,\infty)$ in the points λ_n, what is their growth? Further, to what extent are they unique?

For general sets of interpolation points, the problem of existence and uniqueness of interpolating entire functions of exponential type can be quite tricky and has been studied only up to a certain point. Though it is an interesting problem in itself it is not what we are primarily interested in. In order to avoid having too many complications let us restrict ourselves to the case of equally spaced points. The corresponding questions of uniqueness can be settled with the help of the well known theorem of F. Carlson and its generalizations.

2. Uniqueness Theorems

It was proved by Carlson ([7], also see [5, Chapter 9]) that if g is regular and of exponential type $\tau < \pi$ in the right half plane and $g(n) = 0$ for $n = 1,2,\ldots$, then $g(z) \equiv 0$. From this it follows that an entire function which vanishes at all the positive and negative integers and assumes the value 1 at the origin cannot be of exponential type $< \pi$. No doubt, there are entire functions of exponential type which assume the value 1 at the origin and vanish at the points $\pm 1, \pm 2, \ldots$. As examples we may mention $\frac{\sin(\pi z)}{\pi z} + c \sin(\pi z)$, $\left(\frac{\sin(\pi z)}{\pi z}\right)^2$.

The following lemma which is also well known will help us fix the fundamental functions of Lagrange interpolation.

LEMMA 1. *If g is an entire function of exponential type τ such that*

$$\lim_{x \to \pm\infty} \frac{g(x)}{x} = 0 , \tag{2.1}$$

$g(\frac{\nu\pi}{\tau}) = 0$ *for* $\nu = 0, \pm 1, \pm 2, \ldots$, *then* $g(z) = c \sin \tau z$ *where* c *is a constant.*

This result can be deduced from the following simple consequence of the Phragmén-Lindelöf theorem.

LEMMA 2 [5, Theorem 6.2.4]. *If f is regular and of exponential type in the upper half plane,*

$$h_f(\pi/2) := \limsup_{y \to \infty} \frac{\log|f(iy)|}{y} \le c ,$$

and $|f(x)| \le M$ *for* $-\infty < x < \infty$, *then*

$$|f(x+iy)| \le Me^{cy} , \qquad -\infty < x < \infty,\ 0 \le y < \infty. \tag{2.2}$$

Proof of Lemma 1. Applying Lemma 2 to the function $f(z) = \frac{g(z)-g(0)}{z}$ we obtain

$$g(z) = O((1+|z|)e^{\tau|y|}) \qquad (z = x+iy) . \tag{2.3}$$

Now consider the entire function

$$\varphi(z) = \frac{g(z)}{\sin(\tau z)} .$$

On the boundary of the square

$$S_\nu := \{z = x+iy \in \mathbb{C}: |x - \frac{\nu\pi}{\tau}| \le \frac{\pi}{2\tau}, |y| \le \frac{\pi}{2\tau}\}$$

where ν is an arbitrary integer, we have $|\sin(\tau z)| \ge 1$, and so by (2.3) and the maximum modulus principle

$$(2.4) \qquad |\varphi(z)| = O(1+|z|), \qquad -\infty < \operatorname{Re} z < \infty, \ |\operatorname{Im} z| \le \frac{\pi}{2\tau} .$$

Since $|\sin(\tau z)| \ge (e^{\tau|y|}-e^{-\tau|y|})/2$ it follows from (2.3) that (2.4) holds also for $|y| > \pi/2\tau$. Hence by a generalized version of Liouville's theorem $\varphi(z) = c+dz$, i.e. $g(z) = (c+dz)\sin(\tau z)$. But d must be zero since $\lim_{x\to\infty} \frac{g(x)}{x} = 0$ by hypothesis.

The next lemma will be useful in connection with Hermite interpolation.

LEMMA 3. *If f is an entire function of exponential type 2τ such that $f(\lambda) = f'(\lambda) = 0$ for $\lambda = 0, \pm\frac{\pi}{\tau}, \pm\frac{2\pi}{\tau}, \ldots$, and $|f(x)|$ is bounded on the real axis, then $f(z) = c \sin^2(\tau z)$ where c is a constant.*

<u>Proof</u>. Let $|f(x)| \le M$ for $-\infty < x < \infty$. Then by Lemma 2

$$(2.5) \qquad |f(x+iy)| \le Me^{2\tau|y|}, \qquad -\infty < x < \infty, \ -\infty < y < \infty.$$

In the same way as above we can show that the entire function (of exponential type)

$$F(z) = \frac{f(z)}{\sin^2(\tau z)}$$

satisfies the inequality

$$(2.6) \qquad |F(x+iy)| \le Me^{\pi}, \qquad -\infty < x < \infty, \ |y| \le \frac{\pi}{2\tau} .$$

If $|y| > \frac{\pi}{2\tau}$, then again in view of (2.5)

$$(2.7) \qquad |F(iy)| = \frac{4|f(iy)|}{e^{2\tau y}-2+e^{-2\tau y}} < \frac{4M}{1-2e^{-\pi}} .$$

By Lemma 2 and Liouville's theorem it follows that F is a constant. Hence Lemma 3 holds.

Finally, in connection with (0,2)-interpolation by entire functions of exponential type we will need:

LEMMA 4 [10, Theorem 4]. *Let* α *be an arbitrary number in* [0,1). *If* f *is an entire function of exponential type* 2τ *such that*

(*i*) $|f(x)| \leq A + B|x|^{\alpha}$ *for all real* x *and certain constants* A, B,

(*ii*) $f(\lambda) = f''(\lambda) = 0,\ \lambda = 0, \pm\frac{\pi}{\tau}, \pm\frac{2\pi}{\tau}, \ldots,$

then $f(z) = c_1 \sin(\tau z) + c_2 \sin(2\tau z)$ *where* c_1, c_2 *are constants.*

3. Lagrange Interpolation and Approximation

For a given positive number τ, let us take as interpolation points

$$\lambda_{\tau,\nu} = \frac{\nu\pi}{\tau}, \qquad \nu = 0, \pm1, \pm2, \ldots \tag{3.1}$$

and try to fix the "fundamental functions".

The entire function

$$g_{\tau,n}(z) = \begin{cases} \dfrac{\sin(\tau z - n\pi)}{\tau z - n\pi}\,\dfrac{\tau z}{n\pi} & \text{if } n \neq 0 \\[2ex] \dfrac{\sin(\tau z)}{\tau z} & \text{if } n = 0 \end{cases}$$

(i) is of exponential type τ, (ii) vanishes at all the points (3.1) except $\lambda_{\tau,n}$, (iii) assumes the value 1 at $\lambda_{\tau,n}$, (iv) its derivative vanishes at the origin, and (v) is bounded on the real axis. According to Lemma 1 it is the only entire function satisfying conditions (i) - (v). The functions $g_{\tau,n}$ will be our "fundamental functions of Lagrange interpolation in the points $\lambda_{\tau,n}$". If f is bounded on the real axis then

$$L_{\tau}(f;z,c) := \frac{c}{\tau}\sin(\tau z) + \sum_{n=-\infty}^{\infty} f(\lambda_{\tau,n})\, g_{\tau,n}(z)$$

is an entire function of exponential type τ which interpolates the function f in the points $\lambda_{\tau,n}$ and whose derivative at the origin is equal to c. Moreover, $L_{\tau}(f;x,c) = O(\log|x|)$ as $x \to \pm\infty$ (see Lemma 5 below). According to Lemma 1 $L_{\tau}(f;z,c)$ is uniquely determined by all these properties.

We shall now see that for a given $f \in UCB(-\infty,\infty)$ howsoever we choose the values c_τ the error

$$\sup_{-\infty<x<\infty} |f(x)-L_\tau(f;x,c_\tau)| \tag{3.2}$$

does not necessarily tend to zero as $\tau \to \infty$, even if we take the supremum in (3.2) only on the interval $[-\Delta,\Delta]$ where Δ is an arbitrary but fixed number $> \pi$. Let $C^*(-\infty,\infty)$ denote the Banach space of all functions f which are continuous on $(-\infty,\infty)$ and vanish outside $[-1,1]$. Then

$$f \to L_\tau(f;\cdot,0)$$

may be considered as a bounded linear transformation L_τ from $C^*(-\infty,\infty)$ to the normed linear space $C[-\Delta,\Delta]$ of all continuous functions g on $[-\Delta,\Delta]$ with $\|g\| = \sup_{-\Delta\le x\le\Delta} |g(x)|$. It is easily verified that

$$\|L_\tau\| = \sup_{-\Delta\le x\le\Delta} \sum_{n=-[\frac{\tau}{\pi}]}^{[\frac{\tau}{\pi}]} |g_{\tau,n}(x)| \ge \sum_{n=1}^{[\frac{\tau}{\pi}]} \left|g_{\tau,n}\left(\frac{(2[\tau]+1)\pi}{2\tau}\right)\right|$$

which for large τ is greater than $C \log \tau$ where C (> 0) is independent of τ. Hence by the Banach-Steinhaus theorem there exists a function $\hat{f} \in C^*(-\infty,\infty)$ and so *a fortiori* in $UCB(-\infty,\infty)$ such that

$$\sup_{-\Delta\le x\le\Delta} |\hat{f}(x)-L_\tau(\hat{f};x,0)|$$

does not remain bounded as $\tau \to \infty$. Obviously the same is true for

$$\sup_{-\Delta\le x\le\Delta} |\hat{f}(x)-L_\tau(\hat{f};x,c_\tau)| \tag{3.3}$$

as long as $c_\tau = O(\tau)$. On the other hand

$$\begin{aligned} |L_\tau(\hat{f};\tfrac{\pi}{2\tau},0)| &\le \max_{-1\le x\le 1} |\hat{f}(x)|\cdot \sum_{n=-[\frac{\tau}{\pi}]}^{[\frac{\tau}{\pi}]} |g_{\tau,n}(\tfrac{\pi}{2\tau})| \\ &\le \max_{-1\le x\le 1} |\hat{f}(x)|\cdot\frac{2}{\pi}\left(1 + \sum_{n=1}^{\infty} \frac{1}{n^2}\right) \\ &= \left(\frac{2}{\pi} + \frac{\pi}{3}\right) \max_{-1\le x\le 1} |\hat{f}(x)| \ , \end{aligned}$$

which implies that

$$|\hat{f}(\tfrac{\pi}{2\tau}) - L_\tau(\hat{f};\tfrac{\pi}{2\tau},c_\tau)| \geq \frac{1}{\tau}\,|c_\tau| - (1 + \frac{2}{\pi} + \frac{\pi}{3}) \max_{-1\leq x\leq 1} |\hat{f}(x)|$$

and so if $\tau = o(c_\tau)$ for some sequence of values τ tending to infinity then again (3.3) cannot remain bounded.

As in the case of Lagrange interpolation by polynomials the situation becomes better if we add an appropriate smoothness condition on the function f.

THEOREM 1. *Let $f \in UCB(-\infty,\infty)$ and suppose that the modulus of continuity* $\omega(\delta,f) := \sup_{|t_1-t_2|\leq\delta} |f(t_1)-f(t_2)|$ *of f satisfies*

$$\lim_{\tau\to\infty} (\tfrac{1}{\tau},f)\log\tau = 0 \quad . \tag{3.4}$$

Then for each τ it is possible to choose c_τ (depending on f) such that $L_\tau(f;x,c_\tau)$ tends to f uniformly on all compact subintervals of the real line as $\tau \to \infty$.

The proof depends on the following

LEMMA 5. *There exist constants c_1, c_2 not depending on τ and x such that*

$$\sum_{n=-\infty}^{\infty} |g_{\tau,n}(x)| \leq c_1 + c_2\,\{\log\tau + \log(|x|+2)\} \tag{3.5}$$

for all large τ.

Proof of Lemma 5. Put $\Lambda(x) := \sum_{n=-\infty}^{\infty} |g_{\tau,n}(x)|$. Since $\Lambda(0) = 1$ and $\Lambda(-x) = \Lambda(x)$ we may assume that $x > 0$. Then

$$\Lambda(x) = |\frac{\sin(\tau x)}{\tau x}| + \sum_{\substack{n=-\infty\\ n\neq 0}}^{\infty} |\frac{\sin(\tau x - n\pi)}{\tau x - n\pi}\,\frac{\tau x}{n\pi}|$$

$$\leq 1 + \sum_{\substack{n=-\infty\\ n\neq 0}}^{\infty} |\frac{\sin(\pi X)}{\pi(X-n)}\,\frac{X}{n}| \qquad \text{where } X = \frac{\tau x}{\pi}$$

$$\leq 1 + \frac{2}{\pi} \sum_{n=1}^{\infty} \frac{|\sin(\pi X)|}{|X-n|} \frac{X}{n}$$

$$\leq 1 + \frac{\pi}{3} \qquad \text{if } 0 < X \leq \frac{1}{2} .$$

If $\frac{2m-1}{2} < X \leq \frac{2m+1}{2}$ for some positive integer $m \geq 1$, then

$$\Lambda(x) \leq 1 + \frac{2X}{m} + A_1 + A_2$$

where

$$A_1 = \begin{cases} 0 & \text{if } m = 1 \\ \frac{2}{\pi} \sum_{n=1}^{m-1} \frac{X}{(X-n)n} & \text{if } m \geq 2 \end{cases}$$

and

$$A_2 = \frac{2}{\pi} \sum_{n=m+1}^{\infty} \frac{X}{(n-X)n} .$$

It is clear that for $m \geq 2$

$$A_1 \leq \frac{2}{\pi} \left\{1 + \log(m-1) + \sum_{n=1}^{m-1} \frac{1}{X-n}\right\}$$

$$\leq \frac{2}{\pi} \left\{3 + \log(m-1) + \log(2(m-1)-1)\right\}$$

whereas

$$A_2 = \frac{2}{\pi} \lim_{N\to\infty} \sum_{n=m+1}^{N} \left(\frac{1}{n-X} - \frac{1}{n}\right)$$

$$\leq \frac{4}{\pi} + \frac{2}{\pi} \log m + \frac{2}{\pi} \lim_{N\to\infty} \log\left\{\frac{2(N-m)-1}{N}\right\}$$

$$= \frac{4}{\pi} + \frac{2}{\pi} \log(2m) .$$

Hence (3.5) holds.

Proof of Theorem 1. Let φ_τ be any entire function bounded on the real axis and of exponential type τ. Then

$$|f(x)-L_\tau(f;x,\varphi_\tau'(0))| \le |f(x)-\varphi_\tau(x)| + |\varphi_\tau(x)-L_\tau(f;x,\varphi_\tau'(0))|$$

$$= |f(x)-\varphi_\tau(x)| + |L_\tau(\varphi_\tau;x,\varphi_\tau'(0))-L_\tau(f;x,\varphi_\tau'(0))| \text{ by Lemma 1}$$

$$= |f(x)-\varphi_\tau(x)| + |L_\tau(\varphi_\tau-f;x,0)|$$

$$\le |f(x)-\varphi_\tau(x)| + \Lambda(x) \sup_{-\infty<x<\infty} |\varphi_\tau(x)-f(x)|$$

$$\text{(3.6)} \quad \le (\Lambda(x)+1) \sup_{-\infty<x<\infty} |f(x)-\varphi_\tau(x)| \quad .$$

Now take in particular

$$\text{(3.7)} \qquad \varphi_\tau(x) = \frac{1}{\mu}\int_{-\infty}^{\infty} \left\{\frac{\sin(\tau(x-t)/4)}{x-t}\right\}^4 f(t)\, dt$$

where

$$\mu = \int_{-\infty}^{\infty} \left\{\frac{\sin(\tau t/4)}{t}\right\}^4 dt \quad .$$

Then [16, p. 259] for some constant c_3

$$\sup_{-\infty<x<\infty} |f(x)-\varphi_\tau(x)| < c_3\, \omega(\tfrac{1}{\tau},f) \quad .$$

Hence from (3.6) and Lemma 5 we get

$$|f(x)-L_\tau(f;x,\varphi_\tau'(0))| < \{1+c_1+c_2(\log\tau + \log(|x|+2))\}\, c_3\, \omega(\tfrac{1}{\tau},f)$$

which, in view of (3.4), implies that $L_\tau(f;x,\varphi_\tau'(0))$ converges to f uniformly on all compact subintervals of $(-\infty,\infty)$.

Under a stronger smoothness condition we can show uniform convergence on the whole real line.

THEOREM 2. *Let $f \in UCB(-\infty,\infty)$. If for some $\varepsilon > 0$ the modulus of continuity $\omega(\delta,F_\varepsilon)$ of the function $F_\varepsilon(x) = \{\log(|x|+2)\}^{1+\varepsilon} f(x)$ satisfies*

$$\lim_{\tau\to\infty} \omega(\tfrac{1}{\tau},F_\varepsilon)\log\tau = 0 , \tag{3.8}$$

then $L_\tau(f,x,\varphi_\tau'(0))$ converges to f uniformly on $(-\infty,\infty)$ as $\tau \to \infty$. Here φ_τ is the function defined in (3.7).

For this we need

LEMMA 6. *For every $\varepsilon > 0$ there exists a constant c_4 independent of x and τ such that*

$$\sum_{n=-\infty}^{\infty} \frac{|g_{\tau,n}(x)|}{\left\{\log\left(\left|\frac{\pi n}{\tau}\right|+2\right)\right\}^{1+\varepsilon}} < c_4 \log\tau \tag{3.9}$$

for all large τ.

Proof of Lemma 6. For $|x| < \tau^4$ the desired estimate follows from (3.5). So let $|x| \geq \tau^4$. In addition we may assume x to be positive. If we set $X = \tau x/\pi$ and suppose that $(2m-1)/2 < X \leq (2m+1)/2$, then as in the proof of Lemma 5 we obtain

$$\sum_{n=-\infty}^{\infty} \frac{|g_{\tau,n}(x)|}{\left\{\log\left(\left|\frac{\pi n}{\tau}\right|+2\right)\right\}^{1+\varepsilon}} \leq \frac{1}{(\log 2)^{1+\varepsilon}} + \frac{2}{\pi}\sum_{n=1}^{\infty} \frac{|\sin(\pi X)|X}{|X-n|n\left\{\log\left(\frac{\pi n}{\tau}+2\right)\right\}^{1+\varepsilon}}$$

$$\leq \frac{1}{(\log 2)^{1+\varepsilon}}\left(1+\frac{2}{\pi}\frac{X}{m}\right) + \frac{2}{\pi}\left(\sum_{n=1}^{[\tau^2]} + \sum_{n=[\tau^2]+1}^{m-1} + \sum_{n=m+1}^{\infty}\right)\frac{X}{|X-n|}\,\frac{1}{n\left\{\log\left(\frac{\pi n}{\tau}+2\right)\right\}^{1+\varepsilon}}$$

If $1 \leq n \leq [\tau^2]$, then $\frac{X}{|X-n|} = \frac{X}{X-n} \leq 2$ for all large τ and so

$$\sum_{n=1}^{[\tau^2]} \frac{X}{|X-n|}\,\frac{1}{n\left\{\log\left(\frac{\pi n}{\tau}+2\right)\right\}^{1+\varepsilon}} \leq \frac{2}{\log 2}\sum_{n=1}^{[\tau^2]}\frac{1}{n} = O(\log\tau) .$$

If $[\tau^2]+1 \leq n \leq m-1$, then $\frac{\pi n}{\tau} \geq \sqrt{n}$ and so

$$\sum_{n=[\tau^2]+1}^{m-1} \frac{X}{|X-n|} \frac{1}{n\left\{\log\left(\frac{\pi n}{\tau}+2\right)\right\}^{1+\varepsilon}} \le \sum_{n=[\tau^2]+1}^{m-1} \frac{X}{(X-n)n} \frac{2^{1+\varepsilon}}{(\log n)^{1+\varepsilon}}$$

$$= \sum_{n=[\tau^2]+1}^{m-1} \left(\frac{1}{n} + \frac{1}{X-n}\right) \frac{2^{1+\varepsilon}}{(\log n)^{1+\varepsilon}}$$

$$= O(1) + \sum_{n=[\tau^2]+1}^{m-1} \frac{1}{X-n} \frac{2^{1+\varepsilon}}{(\log n)^{1+\varepsilon}}$$

$$< O(1) + 2^{2+\varepsilon} \sum_{k=1}^{m-1-[\tau^2]} \frac{1}{(2k-1)\{\log(m-1-k)\}^{1+\varepsilon}}$$

$$< O(1) + 2^{2+\varepsilon} \left\{ \sum_{k=2}^{[\frac{m-1}{2}]} \frac{1}{k\{\log k\}^{1+\varepsilon}} + \sum_{k=[\frac{m-1}{2}]+1}^{m-1-[\tau^2]} \frac{1}{k} \right\}$$

$$= O(1).$$

Further

$$\sum_{n=m+1}^{\infty} \frac{X}{|X-n|} \frac{1}{n\left\{\log\left(\frac{\pi n}{\tau}+2\right)\right\}^{1+\varepsilon}} \le \sum_{n=m+1}^{\infty} \frac{X}{(n-X)n} \frac{2^{1+\varepsilon}}{(\log n)^{1+\varepsilon}}$$

$$< 2^{1+\varepsilon} \sum_{n=m+1}^{\infty} \frac{1}{n-X} \frac{1}{(\log n)^{1+\varepsilon}}$$

$$\le 2^{2+\varepsilon} \sum_{k=1}^{\infty} \frac{1}{2k-1} \frac{1}{\{\log(m+k)\}^{1+\varepsilon}}$$

$$= O(1).$$

Hence we have the desired estimate (3.9).

Proof of Theorem 2. Let φ_τ be given by the formula (3.7) and let

$$\Omega_\tau(t) := \frac{1}{\mu}\left\{\frac{\sin(\tau t/4)}{t}\right\}^4 .$$

Then

$$\{\log(|x|+2)\}^{1+\varepsilon}|f(x)-\varphi_\tau(x)|$$

$$\leq \{\log(|x|+2)\}^{1+\varepsilon}\int_{-\infty}^{\infty}\Omega_\tau(t)|f(x+t)-f(x)|dt .$$

Now

$$\{\log(|x|+2)\}^{1+\varepsilon}|f(x+t)-f(x)| \leq |\{\log(|x+t|+2)\}^{1+\varepsilon}f(x+t)$$

$$- \{\log(|x|+2)\}^{1+\varepsilon}f(x)| + |f(x+t)|\ |\{\log(|x+t|+2)\}^{1+\varepsilon}$$

$$- \{\log(|x|+2)\}^{1+\varepsilon}| \leq |\{\log(|x+t|+2)\}^{1+\varepsilon}f(x+t)$$

$$- \{\log(|x|+2)\}^{1+\varepsilon}f(x)| + M|\{\log(|x+t|+2)\}^{1+\varepsilon} - \{\log(|x|+2)\}^{1+\varepsilon}|$$

if $|f(x)|$ is bounded by M on the real axis. Putting

$$F_\varepsilon(x) := \{\log(|x|+2)\}^{1+\varepsilon}f(x)$$

we obtain (see [16, p. 259])

$$\{\log(|x|+2)\}^{1+\varepsilon}|f(x)-\varphi_\tau(x)| \tag{3.10}$$

$$\leq c_5\ \omega(\tfrac{1}{\tau},F_\varepsilon) + c_6\ \omega(\tfrac{1}{\tau},\{\log(|x|+2)\}^{1+\varepsilon}) .$$

As in the proof of Theorem 1 we have

$$|f(x)-L_\tau(f;x,\varphi_\tau'(0))| \leq |f(x)-\varphi_\tau(x)| + |L_\tau(\varphi_\tau-f;x,0)|$$

$$\leq c_3\ \omega(\tfrac{1}{\tau},f)$$

$$+\ \left|\sum_{n=-\infty}^{\infty} \frac{g_{\tau,n}(x)}{\left\{\log\left(\left|\frac{\pi n}{\tau}\right|+2\right)\right\}^{1+\varepsilon}} \left(\varphi_\tau\left(\frac{\pi n}{\tau}\right) - f\left(\frac{\pi n}{\tau}\right)\right) \left\{\log\left(\left|\frac{\pi n}{\tau}\right|+2\right)\right\}^{1+\varepsilon}\right| \ .$$

Hence given that (3.8) holds, the desired conclusion follows on using (3.9), (3.10) and the fact that

$$\omega\left(\frac{1}{\tau},\{\log(|x|+2)\}^{1+\varepsilon}\right) \leq \frac{1+\varepsilon}{\tau} \ .$$

4. Hermite-Fejér Interpolation and Approximation

We again interpolate in the points (3.1) but this time we look for an entire function of exponential type which duplicates the given function in the points $\lambda_{\tau,n}$ and whose derivative assumes prescribed values at these points. In fact, we require the derivative of the interpolating entire function to be zero at these points. As such we will not have to worry about "fundamental functions of the second kind". The entire function

$$h_{\tau,n}(z) = \left\{\frac{\sin(\tau z)}{\tau z - n\pi}\right\}^2$$

(a) is of exponential type 2τ, (b) vanishes at all the points (3.1) except $\lambda_{\tau,n}$, (c) assumes the value 1 at $\lambda_{\tau,n}$, (d) its derivative vanishes at all the points (3.1), and (e) $|h_{\tau,n}(x)| = o(1)$ as $x \to \pm\infty$. Given a function $f \in UCB(-\infty,\infty)$ we use these "fundamental functions of the first kind" to construct the entire function

$$H_\tau(f;z) = \sum_{n=-\infty}^{\infty} f(\lambda_{\tau,n})\ h_{\tau,n}(z)$$

which has the properties:

(i) it is of exponential type 2τ

(ii) it is bounded on the real axis

(iii) it interpolates f in the points $\lambda_{\tau,n}$

(iv) its derivative vanishes at these points

(v) $|H_\tau(f;iy)| = O(|y|^{-1}e^{2\tau|y|})$ as $y \to \pm\infty$.

According to Lemma 3 it is the only entire function having these five properties.

Now we prove the following result which may be compared with the theorem of Fejér mentioned earlier.

THEOREM 3. *If* $f \in UCB(-\infty,\infty)$ *then*

$$\text{(4.1)} \qquad \sup_{-\infty<x<\infty} |f(x)-H_\tau(f;x)| \to 0 \qquad \text{as } \tau \to \infty .$$

Proof of Theorem 3. Let $\omega(t,f)$ be the modulus of continuity of f. Given $\varepsilon > 0$, there exists $\delta > 0$ such that $\omega(\delta,f) < \varepsilon/2$. Using the well known identity

$$\sum_{n=-\infty}^{\infty} \left\{\frac{\sin(\tau z)}{\tau z - n\pi}\right\}^2 \equiv 1 \quad ,$$

we obtain

$$|f(x)-H_\tau(f;x)| = \Big|\sum_{n=-\infty}^{\infty} \left(f(x)-f\left(\frac{n\pi}{\tau}\right)\right) h_{\tau,n}(x)\Big|$$

$$\leq \left(\sum_{\left|\frac{n\pi}{\tau}-x\right|<\delta} + \sum_{\left|\frac{n\pi}{\tau}-x\right|\geq\delta}\right) \left|f(x)-f\left(\frac{n\pi}{\tau}\right)\right| \, |h_{\tau,n}(x)|$$

$$\leq \omega(\delta,f) \sum_{n=-\infty}^{\infty} |h_{\tau,n}(x)| + \frac{2M}{\tau^2} \sum_{\left|\frac{n\pi}{\tau}-x\right|\geq\delta} \frac{1}{\left(x - \frac{n\pi}{\tau}\right)^2}$$

where $M = \sup_{-\infty<x<\infty} |f(x)|$. Thus

$$|f(x)-H_\tau(f;x)| \leq \omega(\delta,f) + \frac{2M}{\tau^2}\left(\frac{2}{\delta^2} + \frac{2\tau}{\pi\delta}\right) < \varepsilon$$

if τ is sufficiently large. Hence the theorem is proved.

5. (0,2)-Interpolation and Approximation

Given a function f bounded on the real axis we wish to construct an entire function of exponential type which interpolates f in the points $\lambda_{\tau,n}$ and whose second derivative assumes prescribed values at these points. For this we first fix the fundamental functions of (0,2)-interpolation namely $A_{\tau,n}$, $B_{\tau,n}$ which are required to satisfy the following properties:

(i) $A_{\tau,n}$, $B_{\tau,n}$ are of exponential type 2τ and are bounded on the real axis,
(ii) $A_{\tau,n}$, $B''_{\tau,n}$ assume the value 1 at the point $\lambda_{\tau,n}$ but vanish at the other points of (3.1),
(iii) $A''_{\tau,n}$, $B_{\tau,n}$ vanish at all the points of (3.1),
(iv) $A'_{\tau,n}(0) = A'''_{\tau,n}(0) = B'_{\tau,n}(0) = B'''_{\tau,n}(0) = 0$.

It turns out that [11]

$$A_{\tau,n}(z) = \begin{cases} \dfrac{\sin(\tau z)}{\tau z} + \dfrac{\sin(\tau z)}{\tau} \displaystyle\int_0^z \frac{1}{\zeta^2}\left(1 - \frac{\sin(\tau\zeta)}{\tau\zeta}\right) d\zeta & \text{if } n = 0 \\[2ex] \dfrac{\tau z}{n\pi}\, \dfrac{\sin(\tau z - n\pi)}{\tau z - n\pi} + \dfrac{\sin(\tau z - n\pi)}{\tau} \displaystyle\int_{-\frac{n\pi}{\tau}}^{z-\frac{n\pi}{\tau}} \frac{1}{\zeta^2}\left(1 - \frac{\sin(\tau\zeta)}{\tau\zeta}\right) d\zeta & \\[2ex] \qquad - \dfrac{\sin(\tau z)}{\pi^3 n^3}\left(1 - \cos(\tau z)\right) & \text{if } n \neq 0 , \end{cases}$$

$$B_{\tau,n}(z) = \begin{cases} \dfrac{\sin(\tau z)}{2\tau} \displaystyle\int_0^z \frac{\sin(\tau\zeta)}{\tau\zeta}\, d\zeta & \text{if } n = 0 \\[2ex] \dfrac{\sin(\tau z - n\pi)}{2\tau} \displaystyle\int_{-\frac{n\pi}{\tau}}^{z-\frac{n\pi}{\tau}} \frac{\sin(\tau\zeta)}{\tau\zeta}\, d\zeta + \dfrac{\sin(\tau z)}{2\tau^2 \pi n}\left(1 - \cos(\tau z)\right) & \\[2ex] & \text{if } n \neq 0 , \end{cases}$$

have these four properties and by Lemma 4 are unique. If $\sum_{n=-\infty}^{\infty} |y''_{\tau,n}| < \infty$, then

$$R_\tau(f;z) = \sum_{n=-\infty}^{\infty} f(\lambda_{\tau,n})\, A_{\tau,n}(z) + \sum_{n=-\infty}^{\infty} y''_{\tau,n}\, B_{\tau,n}(z)$$

is an entire function of exponential type 2τ such that

$$R_\tau(f;\lambda_{\tau,n}) = f(\lambda_{\tau,n}), \qquad n = 0,\pm1,\pm2,\ldots$$

$$R''_\tau(f;\lambda_{\tau,n}) = y''_{\tau,n}, \qquad n = 0,\pm1,\pm2,\ldots .$$

Using the Banach-Steinhaus theorem we can prove:

THEOREM 4. *Given $M > 0$ there exists a function $f \in C^*(-\infty,\infty)$ such that*

$$\limsup_{\tau\to\infty} \max_{-M\le x\le M} \left|f(x)-R_\tau(f;x)\right| = \infty .$$

For (0-2)-interpolation by polynomials in points on $[-1,1]$ the zeros of $(1-x^2)P'_{n-1}(x)$ constitute a more appropriate choice than the Tchebycheff nodes. Hence the points (3.1) which "correspond" to these (Tchebycheff) nodes do not appear to be the most suitable for (0,2)-interpolation by entire functions of exponential type in points on the real line. But we do not know what kind of interpolation points will indeed be better. One of the major difficulties will always be to settle the uniqueness of the fundamental functions.

REFERENCES

1. J. Balász and P. Turán, *Notes on interpolation*. III (*Convergence*), Acta Math. Acad. Sci. Hung., vol. 9 (1958), pp. 195-214.

2. S.N. Bernstein, *Sur l'ordre de la meilleure approximation des fonctions continues par des polynomes de degré donné*, Mémoires de l'Académie Royale de Belgique, (2), vol. 4 (1912), pp. 1-103.

3. S.N. Bernstein, *Sur la limitation des valeurs d'un polynome $P_n(x)$ de degré n sur tout un segment par ses valeurs en $(n+1)$ points du segment*, Izv. Akad. Nauk SSSR. Ser. Mat., vol. 8 (1931), pp. 1025-1050.

4. S.N. Bernstein, *Sur la meilleure approximation sur tout l'axe réel des fonctions continues par des fonctions entières de degré fini*. I, C.R. (Doklady) Acad. Sci. URSS (N.S.), vol. 51 (1946), pp. 331-334.

5. R.P. Boas, Jr., *Entire functions*, Academic Press, New York, 1954.

6. T. Carleman, *Sur un théorème de Weierstrass*, Ark. Mat. Astr. Fys., vol. 20B, No. 4 (1927).

7. F. Carlson, *Sur une classe de série de Taylor*, Thesis, Upsala, 1914.

8. G. Faber, *Über die interpolatorische Darstellung stetiger Funktionen*, Jber. Deutsch. Math. Verein., vol. 23 (1914), pp. 192-210.

9. L. Fejer, *Die Abschätzung eines Polynoms in einem Intervalle, wen Schranken für seine Werte und ersten Ableitungswerte in einzelnen Punkten des Intervalles gegeben sind, und ihre Anwendung auf die Konvergenzfrage Hermitescher Interpolationsreihen*, Math. Z., vol. 32 (1930), pp. 426-457.

10. R. Gervais and Q.I. Rahman, *An extension of Carlson's theorem for entire functions of exponential type*, Trans. Amer. Math. Soc., vol. 235 (1978), pp. 387-394.

11. R. Gervais and Q.I. Rahman, *An extension of Carlson's theorem for entire functions of exponential type*. II, Journal of Mathematical Analysis and Applications (to appear).

12. O. Kiš, *Notes on interpolation* (Russian), Acta Math. Acad. Sci. Hung., vol. 11 (1960), pp. 49-64.

13. J. Marcinkiewicz, *Sur la divergence des polynomes d'interpolation*, Acta Litt. Sci. Szeged, vol. 8 (1936/37), pp. 131-135.

14. W. Rudin, *Real and Complex Analysis*, McGraw-Hill Book Company, New York, 1966.

15. C. Runge, *Über empirische Funktionen und die Interpolation zwischen äquidistanten Ordinaten*, Zeitschrift für Mathematik und Physik, vol. 46 (1901), pp. 224-243.

16. A.F. Timan, *Theory of Approximation of Functions of a Real Variable*, Pergamon Press Inc., New York, 1963.

A SURVEY OF RECENT RESULTS ON OPTIMAL RECOVERY

T.J. Rivlin
THOMAS J. WATSON RESEARCH CENTER
IBM
Yorktown Heights, New York 10598

ABSTRACT: A survey of work in the field of optimal recovery of functions was presented in Micchelli and Rivlin [8] (henceforth referred to as M-R). Our purpose in these three lectures is to reintroduce the notion of optimal recovery and give a selective survey of additional work in that area.

1. AN INTRODUCTION TO OPTIMAL RECOVERY

1.1 Some examples

We introduce the notion of optimal recovery with the following simple example. Let $X = L^{\infty}[0,1]$ (the set of essentially bounded real-valued functions on the inerval $[0,1]$). Let K_n be the subset of X consisting of functions $f \in C^{n-1}[0,1]$ satisfying $|f^{(n-1)}(x)-f^{(n-1)}(y)| \leq |x-y|$, $0 \leq x, y \leq 1$, (so that $f^{(n)}(x)$ exists and satisfies $|f^{(n)}(x)| \leq 1$, a.e. on $[0,1]$) for some given $n \geq 1$. Suppose that in addition to knowing that $f \in K_n$ we also know $(f(x_1),\ldots,f(x_n))$ where $x_1,\ldots,x_n$ are given and satisfy $0 \leq x_1 < x_2 < \ldots < x_n \leq 1$. Suppose $0 \leq t \leq 1$. What is the <u>best possible</u> estimate of $f(t)$ using the sampled values and the fact that $f \in K_n$? More precisely: Let $F = \{(f(x_1),\ldots,f(x_n)): f \in K_n\}$. An <u>algorithm</u>, A, is any function with domain F and range in R (the real line). Then

Badri N. Sahney (ed.), Polynomial and Spline Approximation, 225–245.

$$E(A) = \sup_{f \in K_n} |f(t) - A(f(x_1),\ldots,f(x_n))|$$

is the error of the algorithm A, and

$$E^* = \inf_A E(A)$$

is the <u>intrinsic error</u> in the problem. If $E(A^*) = E^*$ we say that A^* is an <u>optimal</u> algorithm and that it provides an optimal recovery of f(t).

We proceed to solve this optimal recovery problem. The polynomial $Q(x) = (x-x_1)\ldots(x-x_n)/n!$ is in K_n, as is $-Q(x)$. If A is any algorithm then

$$|Q(t) - A(Q(x_1),\ldots,Q(x_n))| \le E(A)$$

and

$$|-Q(t) - A(-Q(x_1),\ldots,-Q(x_n))| \le E(A).$$

Since $Q(x_i) = 0$, $i = 1,\ldots,n$ we have

$$|Q(t) - A(0,\ldots,0)| \le E(A)$$

and

$$|Q(t) + A(0,\ldots,0)| \le E(A)$$

which implies $|Q(t)| \le E(A)$ and hence

$$E^* \ge |Q(t)|.$$

In this typical first stage we have found a lower bound for the intrinsic error. Next we exhibit an algorithm whose error does not exceed the lower bound. Consider the algorithm $B:(f(x_1),\ldots,f(x_n)) \to P(t)$ where P is the interpolating polynomial of degree at most n-1 to the data. Suppose that for any

$f \in K_n$ we have shown that $|f(t)-B(f(x_1),\ldots,f(x_n))| = |f(t)-P(t)| \leq |Q(t)|$. Then $E(B) \leq |Q(t)| \leq E^*$, hence $E^* = |Q(t)|$ and polynomial interpolation is an optimal algorithm. No conceivable method of processing the data can provide error less than $|Q(t)|$.

It remains to show that $|f(t)-P(t)| \leq |Q(t)|$. To this end we employ a typical argument relying on Rolle's theorem. If for some $f \in K_n$ and $t \neq x_i$ we had $f(t) - P(t) = a\,Q(t)$ where $|a| > 1$ then $h(x) = f(x) - P(x) - a\,Q(x)$ has n+1 distinct zeros, $x_1,\ldots,x_n$, t and so $h^{(n-1)}(x) = f^{(n-1)}(x) - c - ax$ (where c is a constant) has two distinct zeros in (0,1), by Rolle's theorem. Let these zeros be u and v, then $h^{(n-1)}(u) - h^{(n-1)}(v) = f^{(n-1)}(u) - f^{(n-1)}(v) - a\,(u-v) = 0$, so that $|f^{(n-1)}(u) - f^{(n-1)}(v)| = |a|\,|u-v|$ contradicting the definition of K_n.

Remark 1. If our object were to recover the function f as an element of X, rather than its value at a given t, then our algorithms are functions from F to X, and our argument above is easily seen to imply that $B'\colon (f(x_1),\ldots,f(x_n)) \to P$ is optimal and the intrinsic error is $\max\{|Q(t)|: 0 \leq t \leq 1\}$. It is natural to inquire further in this case as to which set of sample points yields the smallest intrinsic error, and it is clear that this happens when the x_i are the zeros of the Chebyshev polynomial of degree n.

Remark 2. If we wish to recover $f'(x_j)$, (for some fixed j) then the algorithm $B''\colon (f(x_1),\ldots,f(x_n)) \to P'(x_j)$ is optimal and the intrinsic error is $|Q'(x_j)|$, as Rolle's theorem reveals.

Remark 3. If the number of sampling nodes were less than n, then there would be a polynomial of degree at most n-1 (hence an element of K_n) which vanished at the nodes and was arbitrarily large at t. Thus the intrinsic error would be infinite and the optimal recovery problem trivial. However, if we are given n+r sampling nodes, $0 \le x_1 < \ldots < x_{n+r} \le 1$ with $r \ge 1$ we have an interesting problem whose solution is a direct generalization of the case r = 0. The role of Q is taken by the unique (up to multiplication by -1) perfect spline, q, of degree n having r knots $s_1, \ldots, s_r$ satisfying $0 < s_1 < \ldots < s_r < 1$ (that is: a spline made up of polynomials of degree at most n with simple knots at the s_i whose n^{th} derivative is alternately $\pm M$ in each of the subintervals into which the knots divide [0,1]). having the properties that:

i) $q(x_i) = 0,\ i=1,\ldots,n+r$

and

ii) $|q^{(n)}(x)| \equiv 1\ ,\ x \ne s_i, i=1,\ldots,r.$

It turns out that $E^* = |q(t)|$ and the optimal algorithm is interpolation to the data by a spline of degree n-1 with knots at the abovementioned $s_1, \ldots, s_r$ and evaluation at t. The global recovery of f is again obtained in the same way since the knots are independent of t. A detailed study of this problem can be found in M-R (see also de Boor [2], for how to obtain the knots numerically), but we linger for a moment longer to examine the case n=1. It is clear that the desired perfect spline, q, is, in this case, a polygonal line with slope alternately ±1 which cuts the real axis at $x_1, \ldots, x_{r+1}$. Thus (draw a picture!) $s_i = (x_i + x_{i+1})/2$, i=1,...,r, and an optimal algorithm is obtained as follows: if, say, $s_{j-1} < t < s_j$ ($s_0 = 0$, $s_{r+1} = 1$) estimate f(t) by $f(x_j)$. This case generalizes to higher dimensions (cf. Rivlin [11]). Note, incidentally, that when n=1 the function $|q| \in K_n$ and so we can make the optimal recovery of

$$\int_0^1 f(t)dt$$

by means of

$$\sum_{j=1}^{r+1} (s_j - s_{j-1}) f(x_j).$$

The intrinsic error is

$$\int_0^1 |q(t)|dt.$$

1.2 A general setting

The examples we gave in the previous section have their place in a rather general framework first described in M-R.

Let X be a linear space and K a convex balanced subset of X ("balanced" means $x \in K$ implies $-x \in K$). Let U be a linear operator from X into another normed linear space Z. Our objective is to approximate Ux for $x \in K$ using limited information about x.

A linear operator, I, the information operator, maps X into another normed linear space, Y, and we assume that Ix is known, but only to within a pre-assigned tolerance $\varepsilon \geq 0$. Thus we actually know some $y \in Y$ which satisfies $||y-Ix|| \leq \varepsilon$. An algorithm is any function A whose domain is $IK+\varepsilon S$ (S is the unit ball in Y) and whose range is in Z. The algorithm A produces an error

$$E(A) = \sup_{\begin{cases} x \in K \\ ||Ix-y|| \leq \varepsilon \end{cases}} ||Ux-Ay||.$$

$$E^* = \inf_A E(A)$$

is called the intrinsic error in the problem. If $E(A^*) = E^*$ we say that A^* is an optimal algorithm and that it provides an optimal recovery of Ux. A diagram of the process is given in Fig. 1.

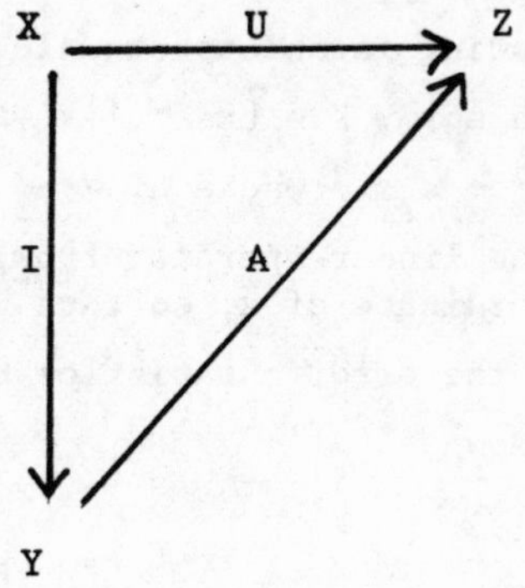

Fig. 1

Thus in the example with which we started $X = L^{\infty}[0,1]$, $K = K_n$, $Uf = f(t)$, $Z = R$, $Y = R^n$, $If = (f(x_1),\ldots,f(x_n))$ and $\varepsilon = 0$.

The lower bound for the intrinsic error which we got in the example is a special case of the following useful general result. (Cf. M-R).

Theorem 1.

$$(1) \quad e(K,\varepsilon) := \sup_{\begin{cases} x \in K \\ ||Ix|| \le \varepsilon \end{cases}} ||Ux|| \le E^*(K,\varepsilon)$$

Proof. Let 0 be the zero vector in Y. If $x \in K$ satisfies $||Ix|| \le \varepsilon$ then for any algorithm A

$$||Ux-A0|| \le E(A) \ ; \ ||U(-x) - A(0)|| \le E(A),$$

according to the definition of E(A). Therefore,

$$||2Ux|| = ||Ux - A0 + Ux + A0|| \le 2E(A)$$

and (1) follows since A was arbitrary.

Remark 4. In our example we actually had equality in (1). That was no accident. Indeed, when Z = R or C (the complex numbers) then with some mild further assumptions on IK there is equality in (1) and, moreover, there exists an optimal algorithm which is linear (i.e., a linear functional on Y). However, equality in (1) does not hold in the full general setting, (details are in M-R) and even though equality holds in (1) there may not be a linear optimal algorithm. Melkman and Micchelli [6] show this last fact with the following ingenious example: Choose $X = R^2$, with the usual euclidean norm, $K = \{x \in X: ||x|| \le 1\}$. Let $Z = R^2$, normed by $||x||_Z^4 = a_1 x_1^4 + a_2 x_2^4$ where $a_1 > a_2 > 0$ are given, and let $Ux = x$ define the linear operator from X into Z. Let Ix produce the first coordinate of x, so that $I(x_1,x_2) = x_1$ and $Y = R$, and suppose that the error ε satisfies $0 < \varepsilon < b/3$ where $b = (2a_2/(a_1+a_2))^{1/2}$. Now

$$(2) \quad e^4(K_1\varepsilon) = \sup_{\begin{cases} x_1^2+x_2^2 \le 1 \\ x_1^2 \le \varepsilon^2 \end{cases}} (a_1x_1^4+a_2x_2^4) = \max(a_2, a_1\varepsilon^4+a_2(1-\varepsilon^2)^2)$$

$$= a_2$$

since $\varepsilon \le b$.

However, every linear algorithm is of the form $Cy = (c_1y, c_2y)$, hence

$$E^4(C) = \max_{\begin{cases} x_1^2+x_2^2 \le 1 \\ |x_1-y| \le \varepsilon \end{cases}} (a_1(x_1-c_1y)^4+a_2(x_2-c_2y)^4).$$

But consideration of $x_1 = y = 1$, $x_2 = 0$ and $x_1 = 0$, $x_2 = 1, y = \pm\varepsilon$ yields

$$E^4(C) \ge \max(a_1(1-c_1)^4+a_2c_2^4,\ a_1(c_1\varepsilon)^4 + a_2(1\mp c_2\varepsilon)^4)$$

$$> a_2.$$

However, there is a non-linear algorithm which is optimal. Namely take $A^*y = 0$ for $|y| \le 2\varepsilon$ and $A^*y = (y,0)$ otherwise. In the first case, $|x_1-y| \le \varepsilon$ implies that $|x_1| \le 3\varepsilon < b$ while in the second we obtain $|x_1| \ge \varepsilon$, hence $x_2^2 \le 1-\varepsilon^2$, and $E^4(A^*) \le a_2$, with the same argument as established (2).

<u>Remark 5</u>. Theorem 1 remains valid if U and I are non-linear transformations satisfying $U(-x) = -Ux$ and $||Ix|| = ||I(-x)||$ for all $x \in X$.

Methods of getting optimal algorithms in general are hard to come by, but the following simple observation is sometimes helpful. Let $K_0 = \{x \in K: Ix = 0\}$.

Theorem 2. (M-R) Suppose there exists a transformation $G: IX \to X$ such that $(x-GIx) \in K_0$ for all $x \in K$, and $\varepsilon = 0$, then $e(K,0) = E(K,0)$ and UG is an optimal algorithm.

Proof.

$$E(UG) = \sup_{x\in K} ||Ux-UGIx|| = \sup_{x\in K} ||U(x-GIx)||$$

$$\leq \sup_{\begin{cases} x\in K \\ Ix=0 \end{cases}} ||Ux|| = e(K,0).$$

As an application of Theorem 2, suppose $n\geq 1$ and X is the space of functions in $C^{n-1}[0,1]$ having an nth derivative in $L^2[0,1]$. Suppose Z = X equipped with the L^2-norm on [0,1], and let Uf = f define the linear operator from X into Z. Suppose $0\leq x_1<\ldots<x_{n+r}\leq 1$ and $If = (f(x),\ldots,f(x_{n+r}))$. Put $K = \{f\in X: ||f^{(n)}||_2\leq 1\}$. We construct the desired G as follows: Let S be the set of natural splines of degree 2n-1 with respect to the knots $x_1,\ldots,x_{n+r}$. Given $f\in X$ let s_f be the unique element of S satisfying

$$s_f(x_i) = f(x_i),\ i=1,\ldots,n+r.$$

Note that $s_f \in X$, and satisfies

$$(3) \qquad ||f^{(n)} - s_f^{(n)}||_2 \leq ||f^{(n)}||_2.$$

Thus if $G: If \to s_f$ then $f-GIf \in K_0$ for all $f\in K$ in view of (3). G satisfies the requirements of Theorem 2 and natural spline interpolation is an optimal algorithm. Note that this remains true of Us_f for any U. Other examples of Theorem 2 may be found in M-R.

2. OPTIMAL RECOVERY FROM INEXACT INFORMATION

In the main example of Section 1.1 we assumed that the sampled function values were accurate. The general setting given in Section 1.2 allows for some error in the information. In this chapter we want to present some results concerning optimal recovery from inexact information which did not appear in M-R.

2.1 An example

We begin by revisiting the example of Section 1.1. A function $f \in K_n$ is sampled at $x_1,\ldots,x_n$, but we know only quantities $y_1,\ldots,y_n$ which satisfy $|f(x_i)-y_i| \le \varepsilon_i$, $i=1,\ldots,n$, while $\varepsilon_i \ge 0$ are given numbers. Our goal, once again, is to estimate $f(t)$ using the available data, $y_1,\ldots,y_n$, this time. Thus the error of an algorithm, A, is now,

$$E(A) = \sup_{\begin{cases} f \in K_n \\ |f(x_i)-y_i| \le \varepsilon_i, i=1,\ldots,n \end{cases}} |f(t)-A(y_1,\ldots,y_n)| \quad .$$

(Note that if we do not require all ε_i to be equal the general setting has to be changed slightly in order to accommodate this example. It is easy to see that Theorem 1, for example, remains valid under this slight alteration.) Suppose that $x_{j-1} \le t \le x_j$. The polynomial of degree at most n-1, p(x), which satisfies

$$p(x_i) = \varepsilon_i \operatorname{sgn} \ell_i(t) \quad , \; i = 1,\ldots,n$$

where ℓ_i are the fundamental polynomials of degree n-1, i.e.,

$$\ell_i(x_j) = \delta_{ij} \; ; \; i,j = 1,\ldots,n,$$

may be written in the form

$$p(x) = \sum_{i=1}^{n} (\varepsilon_i \operatorname{sgn} \ell_i(t))\ell_i(x).$$

Now, $g = Q \operatorname{sgn} Q(t) + p \in K_n$, $|g(x_i)| = \varepsilon_i$, $i = 1,\ldots,n$ and

$$|g(t)| = |Q(t)| + \sum_{i=1}^{n} \varepsilon_i \, |\ell_i(t)|,$$

so that in view of (1), which, as amended, reads

$$E^*(K_n;\varepsilon_1,\ldots,\varepsilon_n) \geq \sup_{\substack{f\in K_n \\ |f(x_i)|\leq\varepsilon_i}} |f(t)|,$$

we see that $E^* \geq |g(t)|$,

But if V is the interpolating polynomial satisfying $V(x_i) = y_i$, $i=1,\ldots,n$, P is as before and $|f(x_i)-y_i|\leq\varepsilon_i$,

$$\begin{aligned} |f(t) - V(t)| &\leq |f(t) - P(t)| + |P(t) - V(t)| \\ &\leq |Q(t)| + \sum_{i=1}^{n} |f(x_i)-y_i| \; |\ell_i(t)| \\ &\leq |Q(t)| + \sum_{i=1}^{n} \varepsilon_i|\ell_i(t)| = |g(t)|. \end{aligned}$$

Thus, polynomial interpolation of the inexact data is an optimal algorithm. Note that the perfect spline g meets the tolerances precisely at the x_i and that its sign configuration on $x_1,\ldots,x_n$ displays alternation except for the permanence at the endpoints of the interval containing t.

If we are given n+r sampling nodes, (r≥1) then the problem with inexact data, does not have as neat a solution as in the case of exact data. However, Micchelli [7] has recently studied this case in some detail and arrived at some interesting conclusions. We shall present next some of his results about the simple case n=1 which together with the example r=0 shed some light on the rather intricate general situation.

Suppose then that we know $y_1,\ldots,y_{r+1}$ satisfying $|y_i-f(x_i)| \leq \varepsilon_i$, $i=1,\ldots,r+1$, where $f\in K_1$. If $0 \leq t \leq 1$ we claim that

$$(3) \quad e(K_1;\varepsilon_1,\ldots,\varepsilon_{r+1}) \geq \min_{i=1,\ldots,r+1} (\varepsilon_i+|t-x_i|).$$

(The equality in (3) is obvious from a picture, but we write down the one line proof anyway).

Let

$$F(x) = \min_{i=1,\ldots,r+1} (\varepsilon_i + |x-x_i|),$$

then $F \in K$, $|F(x_i)| \le \varepsilon_i$, $i=1,\ldots,r+1$ and so $e \ge F(t)$.

Let $T(t) = \{i\colon \varepsilon_i + |t-x_i| = F(t)\}$. If $j \in T(t)$ then $A\colon (y_1,\ldots,y_{r+1}) \to y_j$ is an optimal algorithm. For,

$$|f(t) - y_j| \le | f(t)-f(x_j) | + |f(x_j)-y_j|$$
$$\le |t-x_j| + \varepsilon_j = F(t).$$

Thus we also have equality in (3). Note that the optimal algorithm is linear. Indeed, the general theory given in M-R shows that a linear optimal algorithm must exist, and in the present case a complete description of all linear optimal algorithms can be given. Namely, Micchelli [7] shows:

Theorem 3. $A : R^{r+1} \to R$ is a linear optimal algorithm for recovery of f(t) if, and only if,

$$A(y_1,\ldots,y_{r+1}) = \sum_{i \in T(t)} a_i y_i \ , \ a_i \ge 0 \ , \ \sum_{i \in T(t)} a_i = 1.$$

Note that only data with indices in T(t) need be used in an optimal recovery.

In the general situation Micchelli [7] shows that there is a perfect spline analogous to the q(see Remark 3 of 1.1) of the case $\varepsilon_i = 0$, $i = 1,\ldots,n+r$. It turns out to have the kind of sign configuration suggested by the case $r = 0$. In addition, sample points associated with large errors can be discarded (as the case $n = 1$ suggests) and an optimal algorithm is spline interpolation based on the "acceptable" data. We refer the reader to Micchelli [7] for details.

In the analogous problem when the L^2 norm is used to define K_n and to assign a measure to the data inaccuracy an optimal algorithm requires all of the data. The generalization of the example following Theorem 2 to the case of inaccurate data as well as other optimal estimation problems in Hilbert spaces from inaccurate data are treated in Melkman and Micchelli [6].

2.2 Optimal recovery of a non-linear operator

In all the examples we have presented U was linear. Remark 5 following Theorem 1 suggests that we are not helpless when U is non-linear. We wish to examine such an example next. In the notation of 1.2 let us suppose that B is [-1,1] and $X = C(B)$, the continuous functions on B. Let $K = \{f \in C(B): ||f|| \leq 1\}$. Let P_n denote the polynomials of degree at most n and let $p_n(f)$ be the polynomial of degree at most n which is the best uniform approximation to f on B. We take $Z=C(B)$ and $Uf = p_n(f)$. Thus, U is a non-linear operator which satisfies $U(-f) = -U(f)$. Suppose that $a = \{a_1,\ldots,a_N\}$ is a set of points such that $-1 \leq a_1 < a_2 < \ldots < a_N \leq 1$. Take $Y = R^N$ (equipped with the maximum norm) and let $If = (f(a_1),\ldots,f(a_N))$ for all $f \in C(B)$. Finally suppose that $\varepsilon \geq 0$ is given so that what is available to us in attempting to recover $p_n(f)$ is $y : (y_1,\ldots,y_N)$, where $|y_i - f(a_i)| \leq \varepsilon, i=1,\ldots,N$. In summary, our problem is to recover the best polynomial approximation to f which satisfies $||f|| \leq 1$ from N, possibly inaccurate, sample values of f. This optimal recovery problem is solved in Micchelli and Rivlin [9]. We want to sketch the solution here.

It turns out that a key role in the solution is played by the following polynomial extremal problem. For any real x and $n \geq 1$ determine

$$L_{n,N}(x) = \begin{cases} \sup\limits_{p \in P_n} |p(x)|. \\ |p(a_i)| \leq 1, \; i=1,\ldots,N \end{cases}$$

Note that $L_{n,N}(a_i) = 1$. Suppose henceforth that $x \notin a$. If $N \leq n$ then $q(t) = c(t-a_1)\ldots(t-a_N) \in P_n$ and $|q(x)|$ can be arbitrarily large so that $L_{n,N}(x) = \infty$. Hence we consider only the case $N \geq n+1$. Let $J \subset \{1,\ldots,N\}$ be any set of n+1 distinct integers and let $a(J) = \{a_j : j \in J\}$ be the corresponding subset of a. Put

$$\ell_i(a(J),x) = \frac{\prod\limits_{\substack{j \in J \\ j \neq i}} (x - a_j(J))}{\prod\limits_{\substack{j \in J \\ j \neq i}} (a_i(J) - a_j(J))}, \quad i \in J$$

and let

$$\lambda_n(a(J),x) = \sum_{i \in J} |\ell_i(a(J),x)|$$

be the Lebesgue function for a(J). Micchelli and Rivlin [9] show that

$$L_{n,N}(x) = \min_J \lambda_n(a(J),x).$$

That is: for $N \geq n+1$ the upper envelope of polynomials of degree at most n which are bounded by 1 on the mesh $\underline{a}$ is the lower envelope of the Lebesgue functions of all subsets of $\underline{a}$ consisting of n+1 distinct points. Moreover, it is not hard to see that this lower envelope consists of just one Lebesgue function in each interval of the real line defined by the mesh $\underline{a}$. Thus to each x there is associated a (not necessarily unique) subset J, call it J(x), such that $L_{n,N}(x) = \lambda_n(a(J(x)),x)$, and the same J serves for all x in each interval defined by the mesh $\underline{a}$.

Let

$$L(n,N) = \max_{x \in B} L_{n,N}(x)$$

then Micchelli and Rivlin [9] show that, in the notation of Theorem 1, which is valid here, $e(K,\varepsilon) \geq \min(2,(1+\varepsilon)L(n,N))$. It remains to exhibit an optimal algorithm. To this end we define two algorithms A' and A* by

$$A'y = 0, \text{ all } y \in R^N$$

and if N>n

$$(A^*y)(x) = \sum_{i \in J(x)} y_i \, \ell_i(a(J(x)),x) \quad .$$

We claim that A' is an optimal algorithm if $2 \leq (1+\varepsilon)L(n,N)$ and A^* is an optimal algorithm if $2 > (1+\varepsilon)L(n,N)$.

If $2 \leq (1+\varepsilon)L(n.N)$ then $e(K,\varepsilon) \geq 2$. But since $||f-p_n(f)|| \leq ||f|| \leq 1$, we have $||p_n(f)|| \leq 2$ and so A' is optimal. Suppose then that $2 > (1+\varepsilon)L(n,N)$. Then $N \geq n+1$ and

$$\begin{aligned} ||p_n(f) - A^*y|| &= \max_{x \in B} |p_n(f)(x) - (A^*y)(x)| \\ &= \max_{x \in B} \Big| \sum_{i \in J(x)} [p_n(f)(a_i) - y_i] \ell_i(a(J(x)),x) \Big| . \end{aligned}$$

If we write $p_n(f)(a_i)-y_i = p_n(f)(a_i)-f(a_i) + f(a_i)-y_i$ then $|p_n(f)(a_i)-y_i| \leq 1+\varepsilon$ and hence

$$||p_n(f) - A^*y|| \leq (1+\varepsilon) \max_{x\in B} L_{n,N}(x) = (1+\varepsilon)L(n,N)$$

$$\leq e(K,\varepsilon).$$

This verifies our claim. Note that both optimal algorithms are linear.

In problems of this kind it is natural to ask for the best sampling points, that is, for the set $\underline{a}$ which minimizes the intrinsic error. In the case n = 0,1 the problem is trivial. If N = n+1 we are led to the points having minimal Lebesgue constant which have been characterized by Kilgore, de Boor and Pinkus in their recent work.

3. OPTIMAL RECOVERY IN THE COMPLEX PLANE

We turn now to problems of optimal recovery in the realm of analytic functions. We shall be dealing with the functions of class H^p, $1 \le p \le \infty$, which the reader will recall are defined as follows: If $D := \{x: |z|<1\}$ is the unit disc, a function of $f(z)$ analytic in D is of class H^p, $1 \le p < \infty$ if

$$\left\{ \frac{1}{2\pi} \int_0^{2\pi} |f(re^{i\theta})|^p d\theta \right\}^{\frac{1}{p}}$$

is bounded as $r \to 1$, and of class H^∞ if

$$\max_{0 \le \theta \le 2\pi} |f(re^{i\theta})|$$

is bounded as $r \to 1$. We shall examine first the problem of optimal interpolation of H^∞ functions from exact data, with a few additional comments on the case of inexact data. Then we survey the problem of estimating the integral of H^p functions on the interval $(-1,1)$.

3.1 Optimal interpolation of $\bar{H}^\infty$ functions

In the notation of 1.2 we take $X = H^\infty$ and $K = \{f \in H^\infty: ||f|| \le 1\}$, where

$$||f|| = \sup_{z \in D} |f(z)|.$$

$z_1, \ldots, z_n$ are given points of D, $Y = C^n$ and for each $f \in H^\infty$, $If = (f(z_1), \ldots, f(z_n))$, with the convention that if some of the points $z_1, \ldots, z_n$ coincide the corresponding function values are replaced by derivatives in the obvious way. Suppose that $\zeta \in D$ is given, $Uf = f(\zeta)$ (so that $Z = C$) and $\varepsilon = 0$. Thus we are considering the problem of optimal interpolation of H^∞ functions. The role of the distinguished perfect spline in the real case is taken by the following Blaschke product. Put

$$B_n(z) = \prod_{i=1}^{n} \frac{z-z_i}{1-\bar{z}_i z} .$$

If $f \in K$ and $If=0$ then $f/B_n \in K$ and $|f(\zeta)| \le |B_n(\zeta)|$ by the maximum principle. Therefore

$$e(K,0) = \sup_{\substack{||f|| \le 1 \\ f(z_i)=0, i=1,\dots,n}} |f(\zeta)| = |B_n(\zeta)|.$$

Next define

$$A(f(z_1),\dots,f(z_n)) = \sum_{i=1}^{n} a_i f(z_i)$$

by

$$(4) \qquad f(\zeta) - \sum_{i=1}^{n} a_i f(z_i) = \frac{1}{2\pi i} \int_{|z|=1} \frac{B_n(\zeta)}{B_n(z)} \frac{1-|\zeta|^2}{1-z\bar{\zeta}} \frac{1}{z-\zeta} f(z)dz.$$

Thus for every $f \in K$ we have

$$|f(\zeta) - A(If)| \le |B_n(\zeta)| \frac{1}{2\pi} \int_0^{2\pi} \frac{1-|\zeta|^2}{|e^{i\theta}-\zeta|^2} d\theta ,$$

and since this integral is the Poisson kernel

$$|f(\zeta) - A(If)| \le |B_n(\zeta)| = e(K,0).$$

A is an optimal (linear) algorithm. The a_i are easily computed by the calculus of residues, and the result reveals that A is not polynomial interpolation. For example, if $n=1$ and $z_1=0$, $a_1 = (1-|\zeta|^2)$.

This result as well as the more intricate one when $Uf=f'(\zeta)$ are given in M-R, where relevant literature is also cited. The optimal interpolation problem in H^∞ from inaccurate data seems to be difficult, but D. J. Newman and the author recently analyzed the simplest problem of this kind. Namely, suppose $n=1$, $z_1=0$, and y satisfying $|f(0)-y| \le \varepsilon$ is known. Again we wish to estimate $f(\zeta)$.

We turn first to the intrinsic error. Consider

$$(5) \qquad g(z) = \frac{1}{z} \frac{f(z) - f(0)}{1-\overline{f(0)}f(z)} .$$

Clearly $g \in K$, and

$$f(\zeta) = \frac{\zeta g(\zeta) + f(0)}{1+\zeta g(\zeta)\overline{f(0)}} .$$

Thus, if $\varepsilon < 1$,

$$e = \max_{\substack{||f||\le 1 \\ |f(0)|\le\varepsilon}} |f(\zeta)| = \frac{|\zeta|+\varepsilon}{1+\varepsilon|\zeta|} ,$$

the maximum being attained by, say,

$$f(z) = \frac{z\overline{(\operatorname{sgn}\zeta)}+\varepsilon}{1+\varepsilon\overline{(\operatorname{sgn}\zeta)}z} ,$$

(where $\operatorname{sgn} z = z/|z|$, $z\neq 0$ and $\operatorname{sgn} 0 = 0$. $\overline{\operatorname{sgn} z} = \operatorname{sgn} \bar{z}$). If $\varepsilon\ge 1$, $e = 1$, $f\equiv 1$ is a worst function and zero is an optimal estimate for $f(\zeta)$.

If $\varepsilon<1$ consider the algorithm

(6) $$A: y \to \frac{1-|\zeta|^2}{(1+\varepsilon|\zeta|^2)} y.$$

We claim that this is an optimal algorithm. To see this let us put $w = f(\zeta)$ and $a = f(0)$ for short. Since $g(z)$ as given in (5) satisfies $|g(\zeta)|\le 1$ we have

$$\frac{|w-a|}{|1-\bar{a}w|} \le |\zeta| ,$$

which is equivalent to

(7) $$|w-w_0| \le \rho$$

where

$$w_0 = a \frac{1-|\zeta|^2}{1-|a|^2|\zeta|^2}$$

and

$$\rho = |\zeta| \frac{1-|a|^2}{1-|a|^2|\zeta|^2} .$$

This view of the disc of possible values of $f(\zeta)$ for $f\in K$ helps us verify that (6) defines an optimal algorithm. We have

$$(8)\qquad \left|w - \frac{1-|\zeta|}{(1+\varepsilon|\zeta|)^2}\, y\right| \le \left|w - \frac{1-|\zeta|^2}{(1+\varepsilon|\zeta|)^2}\, a\right| + \frac{1-|\zeta|^2}{(1+\varepsilon|\zeta|)^2}\,\varepsilon.$$

Since w must satisfy (7) we see that the right hand side of (8) does not exceed its value when $w = w_0 + \rho \operatorname{sgn} a$ (or ρ if $a = 0$) which is

$$\frac{|a|+|\zeta|}{1+|a||\zeta|} + \frac{(1-|\zeta|^2)(\varepsilon-|a|)}{(1+\varepsilon|\zeta|)^2}.$$

It remains to show that this last quantity does not exceed $(|\zeta|+\varepsilon)/(1+\varepsilon|\zeta|)$, or that

$$\frac{(1-|\zeta|^2)(\varepsilon-|a|)}{(1+\varepsilon|\zeta|)^2} \le \frac{|\zeta|+\varepsilon}{1+\varepsilon|\zeta|} - \frac{|\zeta|+|a|}{1+|a||\zeta|} = \frac{(\varepsilon-|a|)(1-|\zeta|^2)}{(1+\varepsilon|\zeta|)(1+|a||\zeta|)},$$

which clearly holds.

Note that the linear algorithm given in (6) reduces to that given by (4) when $\varepsilon=0$. (6) gives the unique linear optimal algorithm. There are many non-linear optimal algorithms as examination of the implications of (7) reveals

3.2 Numerical integration of H^p functions

Let us take $X = H^p$, $1 \le p \le \infty$,

$$||f||_p := \lim_{r\to 1} \left[\frac{1}{2\pi}\int_0^{2\pi} |f(re^{i\theta})|^p d\theta\right]^{\frac{1}{p}}.$$

$$||f||_\infty := ||f||.$$

Choose $K = \{f\in H^p : ||f||_p \le 1\}$. $w: (w_1,\dots,w_n)$ are given points of D, $Y = C^n$ and for each $f\in H^p$ $If = (f(w_1),\dots,f(w_n))$. Choose $Z = C$,

$$Uf = \int_{-1}^{1} f(x)dx$$

and $\varepsilon=0$. The intrinsic error in this problem is

$$e(p,w) = \sup_{\substack{||f||_p\le 1\\ f(w_i)=0,\, i=1,\dots,n}} \left|\int_{-1}^{1} f(x)dx\right| .$$

If we put

$$B_n(Z) = \prod_{i=1}^{n} \frac{z-w_i}{1-\overline{w}_i z}$$

then we have

$$(9) \qquad e(p,w) = \sup_{||f||_p \le 1} \left| \int_{-1}^{1} f(x) B_n(x)\,dx \right| .$$

If we ask for the best sampling points, in addition, then we seek

$$e_n(p) = \inf_{w} e(p,w)$$

or equivalently

$$(10) \qquad e_n(p) = \inf_{B_n} \sup_{||f||_p \le 1} \left| \int_{-1}^{1} f(x) B_n(x)\,dx \right| .$$

Given w we know that there exists a linear optimal algorithm, that is, there are constants $c_1(w),\ldots,c_n(w)$ such that

$$\sup_{||f||_p \le 1} \left| \int_{-1}^{1} f(x)\,dx - \sum_{i=1}^{n} c_i f(w_i) \right| \le e(p,w).$$

Bojanov [1] solved this problem when (in our notation) $p = \infty$, K consists of functions which are real on [-1,1] and satisfy $||f|| \le 1$, and there are 2n sample points $w_1, w_1, w_2, w_2, \ldots, w_n, w_n$, that is the function and its derivative are sampled at n given points. Note that in this case (9), immediately yields

$$(11) \qquad e(\infty,w,w') = \int_{-1}^{1} B_n^2(x)\,dx.$$

Bojanov [1] determines the c_i explicitly. He also notes that if w^* is a set of points for which (11) is minimum (the existence of w^* is shown by Loeb [4]) then the coefficients of the derivative evaluations in the corresponding numerical integration formula are all zero. Since (9) implies that for every w

$$e(\infty,w) \ge \int_{-1}^{1} B_n^2(w,x)\,dx$$

we see that

$$e(\infty,w^*) = \int_{-1}^{1} B_n^2(w^*,x)\,dx.$$

The determination of $e(\infty,w)$ in general does not seem to be easy. Boyanov [3] also shows that

$$\exp(-(2\sqrt{2} + \frac{1}{\sqrt{2}})\pi\sqrt{n}) \le e_n(\infty) \le \exp(-\frac{\pi}{\sqrt{2}}\sqrt{n}) .$$

Loeb and Werner [5] showed that if $p^{-1} + q^{-1} = 1$ then

$$e_n(p) \le 2^{1+\frac{2}{q}} \exp(-\frac{1}{2q}([\frac{n}{2}])^{\frac{1}{2}}).$$

A key role in their approach is played by Newman's famous rational approximation to $|x|$. (Cf. Newman [10]).

Stenger [12] recently proved that given $\varepsilon>0$ there exists $n(\varepsilon)$ such that for $n>n(\varepsilon)$

$$\exp(-(5^{\frac{1}{2}}\pi+\varepsilon)n^{\frac{1}{2}}) \le e_n(p) \le \exp(-(\frac{\pi}{2q^{1/2}}-\varepsilon)n^{\frac{1}{2}}).$$

A definitive estimate of $e_n(p)$ was given by D. J. Newman in one of his lectures at the N.S.F. sponsored Regional Conference on Approximation Theory held at the University of Rhode Island, Kingston, Rhode Island, June 12-16, 1978. His lectures will be published in the Regional Conference Series in Applied Mathematics. Newman's result is

$$(12) \quad \frac{2}{3}\exp(-6(\frac{n}{q})^{\frac{1}{2}}) \le e_n(p) \le 11\exp(-\frac{1}{2}(\frac{n}{q})^{\frac{1}{2}}).$$

Note the implication that there is no effective numerical integration formula when p=1. Newman views $e_n(p)$ as given by (10). He obtains his upper bound by defining an explicit Blaschke product by a modification of his well-known rational function. For the lower bound he associates to each Blaschke product, $B_n(x)$, an $f\in K$ which makes

$$\int_{-1}^{1} f(x)B_n(x)\,dx$$

large. This is done by choosing f to be an appropriate multiple of $\overline{B_n(z)})$. For details the reader should consult the lectures when they appear.

REFERENCES

1. Bojanov, B. D., Best quadrature formula for a certain class of analytic functions, Zastos. Mat. 14 (1974), 441-447.

2. de Boor, C., Computational aspects of optimal recovery, in "Optimal Estimation in Approximation Theory", (eds. C. A. Micchelli and T. J. Rivlin), Plenum Press, N. Y., 1977, pp. 69-91.

3. Boyanov, B. D. (= Bojanov, B.D.), Optimal rate of integration and ε-entropy of a class of analytic functions, Mathematical Notes 14 (1973), pp. 551-556. (Trans. from Russian).

4. Loeb, H. L., A note on optimal integration in H_∞, C. r. Acad. Bulgare Sci. 27 (1974), 615-618.

5. Loeb, H. L. and H. Werner, Optimal quadrature in H_p spaces, Math. Z. 138 (1974), 111-117.

6. Melkman, A. A. and C. A. Micchelli, Optimal estimation of linear operators in Hilbert spaces from inaccurate data, IBM Research Report, RC 7175, 1978.

7. Micchelli, C. A., Optimal estimation of smooth functions from inaccurate data, IBM Research Report, RC 7024, 1978.

8. Micchelli, C. A., and T. J. Rivlin, A survey of optimal recovery, in "Optimal Estimation in Approximation Theory", (eds. C. A. Micchelli and T. J. Rivlin), Plenum Press, N. Y., 1977, pp. 1-54.

9. Micchelli, C. A. and T. J. Rivlin, Optimal recovery of best approximations, IBM Research Report, RC 7071, 1978.

10. Newman, D. J., Rational approximation to $|x|$, Michigan Math. J. 11 (1964), 11-14.

11. Rivlin, T. J., Some aspects of optimal recovery, IBM Research Report, RC 6755, 1977.

12. Stenger, F., Optimal convergence of minimum norm approximations in H^p, Numer. Math. 29 (1978), 345-362.

AN INTRODUCTION TO NON-LINEAR SPLINES

Helmut Werner

Institut für Numerische und Instrumentelle Mathematik
und Rechenzentrum
der Westfälischen Wilhelms Universität
Münster/Westfalen, Germany

Abstract:

We introduce functions termed non-linear splines because they belong to a class of k-times differentiable funtions ($k \in \mathbb{N}$) and their restrictions to certain subintervals (defined by knots) are non-linearly dependent on the parameters in contrast to the splines usually considered. It is shown that use of these non-linear splines in interpolation, approximation and numerical quadrature or ordinary differential equations is widely parallel to the linear case as far as numerical work and stability is concerned, while it allows us to take into account special properties of the functions which the spline is used to replace.

§1 Historical Remarks and Examples

The use of spline functions dates back at least to the beginning of this century. Piecewise linear functions had been used already in connection with Peano's existence proof for solutions to the initial value problem of ordinary differential equations, although these functions were not called splines. Splines were first identified in the work of Schoenberg, Sard, and others. Numerous generalizations of splines have been introduced, but the families of functions used in these extensions have always been linear with respect to the free parameters. Of course this allows for the use of functions non-linear in x, i.e., there may be linear combinations of terms like $(x - d_i)^{-1}$, but the numbers d_i would be fixed.

For the approximation of functions where the location of

Badri N. Sahney (ed.), Polynomial and Spline Approximation, 247-306.

the poles is not a priori known it would be much better (and this is verified in the applications) to have the quantities d_i as parameters too, i.e., to use a non-linear family of splines.

One might expect that the construction of such functions will lead to unsurmountable difficulties but we will show that this is not true, and that the numerical methods proceed almost along the same lines as do applications of linear splines, if appropriate non-linear classes are selected. As we go along we are naturally guided to certain axioms, such as the regularity of a spline.

A first step in this direction was taken in the dissertation of Schaback [9]. He investigated *special rational spline functions*. We will use this case as our "*standard example*". We denote $\pi = \{x_1 < \dots < x_m\}$ and $x_0 = \alpha < x_1$, $x_m = \beta$, m = number of subintervals and $I = [\alpha,\beta]$, $I_i = [x_{i-1},x_i]$, $h_i = x_i - x_{i-1}$; then

(1.1) $$S(\pi) = \{s \,|\, s|_{I_i} = a_i + b_i x + \frac{c_i}{d_i - x} \text{ for } i = 1,\dots,m,\ s \in C^2\}$$

is the class of *special rational spline functions*.

Notation. We will use *difference quotients*. Suppose $f(x)$ is pointwise defined. Then

$$\Delta_t^0(x_i)f(t) = f_i = f(x_i),$$ [1)]

(1.2) $$\Delta_t^1(x_i,x_k)f(t) = \frac{f_i - f_k}{x_i - x_k},$$

$$\Delta_t^m(x_0,\dots,x_m)f(t) = \Delta_t^1(x_0,x_m)\cdot\Delta_\tau^{m-1}(t,x_1,\dots,x_{m-1})f(\tau)$$

$$= \sum_{i=0}^{m} f_i \prod_{\substack{j=0\\ j\neq i}}^{m} (x_i - x_j)^{-1}.$$

The last relation is due to the fact that the mth order difference quotient may also be defined as the highest order coefficient of

---

1) This notation is used throughout this article: $f_i' = f'(x_i)$, $s_i'' = s''(x_i)$ etc.

the mth order polynomial interpolating the data (x_i, f_i), $i = 0,\ldots,m$.

For details and proofs compare the quoted text by Werner-Schaback [19].

Remark 1.1: If $f \in C^2$ in a neighborhood of x_i, then we may define

(1.3)
$$\Delta^2(x_i,x_i,x_{i\pm1})f = \lim \frac{1}{y - x_{i\pm1}} [\Delta^1(y,x_i)f - \Delta^1(x_i,x_{i\pm1})f] = [f^1(x_i) - \Delta^1(x_i,x_{i\pm1})f]/(x_i - x_{i\pm1}).$$

An addition furnishes a formula for rewriting a 2nd order difference quotient as the sum of two confluent difference quotients that use data from one subinterval only, i.e.

(1.4)
$$(x_{i+1} - x_{i-1})\Delta^2(x_{i-1},x_i,x_i)f = (x_i - x_{i-1}) \cdot \Delta^2(x_i,x_i,x_{i+1})f + (x_{i+1} - x_i)\Delta^2(x_i,x_i,x_{i+1})f.$$

Remark 1.2: If $r \in R_{2,1}$, i.e. $r = a + bx + \frac{c}{d - x}$, then

(1.5)
$$\Delta^1(x_i,x_{i+1})r = b + \frac{c}{(d-x_i)(d-x_{i+1})},$$

since

$$\left(\frac{1}{d-x_i} - \frac{1}{d-x_{i+1}}\right)/(x_i - x_{i+1}) = \frac{x_i - x_{i+1}}{(d-x_i)(d-x_{i+1})(x_i - x_{i+1})}.$$

The above definition immediately yields

(1.6)
$$\Delta^2(x_{i-1},x_i,x_{i+1})r = \frac{c}{(d-x_{i-1})(d-x_i)(d-x_{i+1})},$$

and in general

$$\Delta^n(x_0,\ldots,x_n)r = c \Big/ \prod_{j=0}^{n} (d - x_j),$$

$n \geq 2$.

In particular,

$$(1.7) \qquad r'' = \frac{2\cdot c}{(d - x_j)^3}\,,\quad r''' = \frac{6\cdot c}{(d - x_j)^4} \text{ at } x = x_j,$$

$$\text{hence} \qquad d - x_j = \frac{3r''(x_j)}{r'''(x_j)}\,,\quad c = \frac{r''}{2}(d - x_j)^3 = \frac{27}{2}\frac{(r'')^4}{(r''')^3}\,.$$

Hence r'', r''' or two different 2nd divided differences determine c, d, and hence every higher order difference and differential quotient.

Remark 1.3: If $c \neq 0$, i.e., if r in Remark 1.2 is not just a linear function, it is seen from (1.6) and (1.7) that the second (and higher) order derivatives of r cannot change sign in the interval of continuity.

§2 The Interpolation Problem

The elements s of $S(\pi)$ are determined by 4 parameters in each subinterval, i.e., there are $4m$ parameters. These parameters are, however, not independent. The continuity requirement $s \in C^2$ adds 3 equations at each interior knot, i.e., $3(m - 1)$ conditions. Therefore, only $m + 3$ parameters are, hopefully, free for adjustment to prescribed conditions, i.e., for the interpolation of given data (x_i, f_i), $i = 0,\ldots,m$. Since there are only $m + 1$ points, two more conditions could (and should) be imposed.

Statement of problem:

P_1: Find $s \in S(\pi)$ such that

(2.1)

1) $s(x_i) = f_i$ for $i = 0,\ldots,m$ ("*Interpolation data*"),

2) $R_0(s) = R_0 f$ ("*Boundary data*"),

$R_m(s) = R_m f$

where R_0 stands for $\frac{d}{dx}$ or $\frac{d^2}{dx^2}$, and the same holds for R_m.

One could try to prescribe $\frac{d}{dx}s$ and $\frac{d^2}{dx^2}s$ at one point, say x_0. Then the interpolation problem could be solved first for $[x_0, x_1]$, then $[x_1, x_2]$ etc. This procedure would amount to solving some (degenerate form of) initial value problem. We will see later, that this leads to a numerically unstable solution and hence should not be used for practical purposes. From Remark 1.3

above and continuity requirements across the knots we obtain

<u>Theorem 2.1</u>: A necessary condition for P_1 to have a solution is

(2.2) $$\operatorname{sgn} \Delta^2(x_{i-1},x_i,x_{i+1})f = \sigma$$

where $\sigma \in \{1,0,-1\}$ independent of i,

$x_{-1} = x_0$ and $i = 0,\dots$ if $R_0 = \frac{d}{dx}$,

$\operatorname{sgn} R_0 f = \sigma$ and $i = 1,\dots$ if $R_0 = \frac{d^2}{dx^2}$,

and

$x_{m+1} = x_m$ and $i = \dots,m$ if $R_m = \frac{d}{dx}$,

$\operatorname{sgn} R_m f = \sigma$ and $i = \dots,m-1$ if $R_m = \frac{d^2}{dx^2}$. □

<u>Notation</u>: Two zeros z_1,z_2 of a function $g(x)$, say, are said to be separated

$$z_1 < z_2 \ \underline{\text{if}}\ \exists\, z \in (z_1,z_2) \text{ with } g(z) \neq 0.$$

It is easily seen that the well known theorem of Rolle holds if "zero" is sharpened to "separated zeros" in its statement.

<u>Lemma 2.1</u>: The second derivative of the difference of two rational splines cannot have two separated zeros in any subinterval I_j.

<u>Proof</u>: Let $v = s - s^*$, then $v'' = \frac{2c}{(d-x)^3} - \frac{2c^*}{(d^*-x)^3}$ in I_j, say, where the coefficients of s^* are marked by a star,

$$v''(z_i) = 0 \Rightarrow \frac{d - z_i}{\sqrt[3]{2c}} = \frac{d^* - z_i}{\sqrt[3]{2c^*}},$$

and if this is true for z_1 and z_2, $z_1 \neq z_2$, both $\in I_j$, then

$$\frac{z_2 - z_1}{\sqrt[3]{2c}} = \frac{z_2 - z_1}{\sqrt[3]{2c^*}},$$

hence $c = c^*$, and $d = d^*$.

Therefore $v'' = 0$.

Immediate Consequence of Lemma 1:

<u>Theorem 2.2 (Unicity)</u>: The solution of P_1 is unique (if it exists).

<u>Proof</u>: Let the splines s_1, s_2 be solutions of P_1. Then $u = s_1 - s_2$ has $m + 1$ zeros in I, hence u' has m interior zeros in I by the theorem of Rolle. u'' has $m - 1$ interior zeros in I by the same theorem. There are two additional zeros by the given boundary data. If $R_0 = \frac{d}{dx}$, an additional induced zero, if $R_0 = \frac{d^2}{dx^2}$ an additional given zero of u'' will arise.

Location of zeros
x given, 0 induced zero

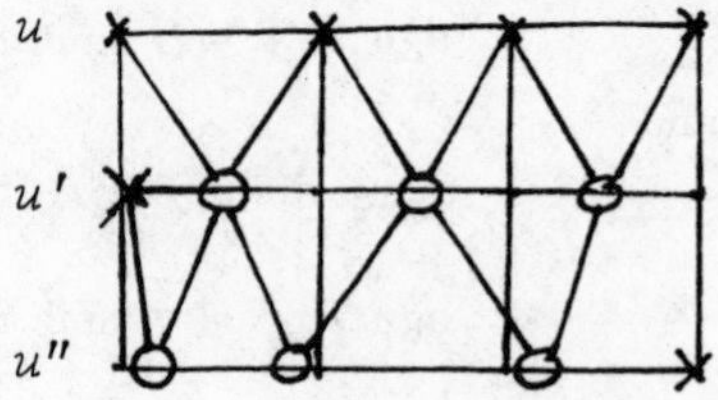

Induction with respect to m:

If $m = 1$, statement follows from Lemma 2.1 and $u(x_0) = u(x_1) = 0$. If $m > 0$, there are $m + 1$ zeros in m subintervals hence there is I_j containing two zeros of u''; these cannot be separated $\Rightarrow u'' \equiv 0 \Rightarrow u \equiv 0$ in I_j by $u(x_{j-1}) = u(x_j) = 0$. Now $(I_1 \cup \ldots \cup I_{j-1})$ and $(I_{j+1} \cup \ldots \cup I_m)$ are intervals with less than m subintervals and at x_{j-1}, respectively x_j, we have the boundary conditions $u'_{j-1} = u''_{j-1} = 0$ and $u'_j = u''_j = 0$, hence the induction hypothesis is applicable. □

<u>Remark 2.1</u>: No special property of s is used except the one used in the verification of Lemma 2.1.

§3. <u>Existence of Interpolating Rational Splines</u>

Consider s restricted to I_j. With the notation $z = x - x_j$ we also have

$$s(x) = \overline{a}_j + \overline{b}_j z + \frac{1}{2}\,\overline{c}_j\,\frac{z^2}{1 - d_j z}, \tag{3.1}$$

and the coefficients satisfy the following relationships:

$$\begin{aligned} \bar{a}_j &= s(x_j), \\ \bar{b}_j &= s'(x_j), \\ \bar{c}_j &= s''(x_j), \\ \bar{c}_j d_j &= \frac{1}{3} s'''(x_j). \end{aligned} \tag{3.2}$$

The last relation follows from the expansion

$$\frac{1}{2}\bar{c}_j z^2/(1 - d_j z) = \frac{1}{2}\bar{c}_j z^2 + \frac{1}{2}\bar{c}_j d_j z^3 + \cdots .$$

From

$$\begin{aligned} \frac{z^2}{1 - d_j z} &= \frac{1}{d_j^2}\left[\frac{1}{1 - d_j z} + \frac{(d_j z)^2 - 1}{1 - d_j z}\right] \\ &= \frac{1}{d_j^2}\left[\frac{1}{1 - d_j z} - 1 - d_j z\right], \end{aligned}$$

we also get

$$s''(x_{j-1}) = \frac{\bar{c}_j}{(1 + d_j h_j)^3} . \tag{3.3}$$

This shows that $\bar{c}_j$ and d_j are easily found from $s''(x_{j-1})$ and $s''(x_j)$ and vice versa. Furthermore $s(x_{j-1})$ and $s(x_j)$ determine $\bar{a}_j$, $\bar{b}_j$:

$$-\bar{b}_j \cdot h_j = s(x_{j-1}) - \bar{a}_j - \frac{1}{2}\bar{c}_j \frac{h_j^2}{1 + d_j h_j} .$$

Hence, to determine the interpolating spline s, all we need to calculate are the values of $s''(x_j)$. If $s''(x_j)$ and $s''(x_{j-1})$ have the same sign ($\neq 0$) from (3.2) and (3.3) we obtain

$$1 + d_j h_j = \sqrt[3]{\frac{s''(x_j)}{s''(x_{j-1})}} ,$$

from which we conclude that $1 - d_j z$ cannot vanish in $[-h_j, 0]$, hence $s(x)$ is continuous in I_j.

Applying a 2nd order difference quotient to (3.1) will annihilate the linear part of s, provided only data from one subinterval are used. This is the reason for rewriting the difference quotient as a sum of confluent difference quotients, given by (1.4). Let

$$\lambda_i = \frac{x_i - x_{i-1}}{x_{i+1} - x_{i-1}} = \frac{h_i}{h_i + h_{i+1}},$$

$$\mu_i = \frac{x_{i+1} - x_i}{x_{i+1} - x_{i-1}} = \frac{h_{i+1}}{h_i + h_{i+1}},$$

thus

$$\lambda_i + \mu_i = 1;$$

then (1.4) yields the equations

$$(3.4) \qquad \lambda_i \Delta^2(x_{i-1}, x_i, x_i)s + \mu_i \Delta^2(x_i, x_i, x_{i+1})s$$

$$= \Delta^2(x_{i-1}, x_i, x_{i+1})f = \Delta_i^2 f,$$

which facilitates elimination of the coefficients $\bar{a}_j$, $\bar{b}_j$. By (1.6) we have

$$(3.5) \qquad \Delta^2(x_{i\pm1}, x_i, x_i)s = \frac{c}{(d - x_{i\pm1})(d - x_i)^2}, \quad \begin{aligned} c &= c_{i+1} \text{ or } c_i, \\ d &= d_{i+1} \text{ or } d_i, \end{aligned}$$

$$= \frac{1}{2} m_i^2 \cdot m_{i\pm1},$$

where $m_i = \dfrac{\sqrt[3]{2c}}{d - x_i} = \sqrt[3]{s''(x_i)}$.

This leads to the "determining equation" for the second order derivatives. All previous considerations are summarized in

<u>Theorem 3.1</u>: Given P_1, let m_i ($i = 0,\ldots,m$) denote the solution of

$$(3.6) \qquad \lambda_i m_{i-1} m_i^2 + \mu_i m_i^2 m_{i+1} = 2\Delta_i^2 f, \quad i = 1,\ldots,m-1,$$

$$\lambda_i = \frac{h_i}{h_i + h_{i+1}}, \quad \mu_i = \frac{h_{i+1}}{h_i + h_{i+1}},$$

and assume

$\operatorname{sgn} m_i = \sigma$, independent of i.

If R_0, R_m equal $\frac{d^2}{dx^2}$ then $m_0 = f''_0$, respectively $m_m = f''_m$, else use $x_{-1} = x_0$ respectively $x_{m+1} = x_m$ and interpret (3.6) appropriately. Then there is a spline $s \in S(\pi)$ that solves P_1. Its construction is described by (3.1), (3.2), (3.3).

How can one prove that (3.6) has a solution? Schaback [10] noted that this system can be written as the gradient of a function of $(m_0, \ldots, m_m)$. We change his notation only slightly. Introduce $z_i = \frac{1}{m_i}$ and multiply (3.6) by $h_i + h_{i+1}$, then we get

$$(3.7) \qquad -E_i = h_{i+1} \frac{1}{z_1^2} \cdot \frac{1}{z_{i+1}} + h_i \frac{1}{z_i^2} \cdot \frac{1}{z_{i-1}} - \underbrace{2(h_i + h_{i+1})\Delta_i^2 f}_{A_i} = 0.$$

This looks like the *gradient of the function*

$$(3.8) \qquad E = \sum_{i=0}^{m} \left(\frac{h_i}{z_i \cdot z_{i-1}} + A_i z_i \right),$$

where $h_0 = 0$.

If $A_i = 0$, a solution of (3.6) is given by $m_i = 0 \ \forall i$. Assume $A_i > 0 \ \forall i$. (Necessary condition is given in Theorem 2.1). Then one may easily establish existence of a minimum of E with every $z_i > 0$.

<u>Proof</u>: Select one set of z_i, for example, $z_i = 1 \ \forall i$, let $E(1, \ldots, 1) = E_1$. If $z_i \geq \frac{E_1}{\min_i A_i} = \bar{z}$ for at least one i then $E(z) \geq E_1$. If all components $\leq \frac{E_1}{\min A_i} = \bar{z}$ and at least one

$$z_i \leq \underline{z} = \frac{\min h_i}{E_1 \cdot \bar{z}},$$

then

$$E(z) \geq \frac{h_i}{z_i z_{i+1}} + A_i z_i \geq \frac{h_i}{\bar{z} \cdot \underline{z}} \geq E_1.$$

Hence we may restrict our attention to the compact domain

$$0 < \underline{z} \leq z_i \leq \overline{z}$$

for all i under consideration (depending upon the boundary conditions).
We thus proved

Theorem 3.2: The necessary conditions given in Theorem 2.1 are sufficient for existence of an interpolating spline s to P_1.

Remark 3.1: $E(z)$ is a convex functional in $z_i > 0$, hence the solution is unique - a new proof of unicity.

Remark 3.2: Another definition of special classes of non-linear splines uses the dual optimization problem to the above-one to define the splines (See Baumeister [2]).

Consider all functions in $C^2(I)$ such that

$$(3.9) \qquad V(f) = \{g \in C^2(I) \mid g(x_i) = f_i \text{ for } i = 0,\ldots,m; \quad R_0 g = R_0 f;\ R_m g = R_m f\}$$

and minimize

$$(3.10) \qquad \int_I A(y'')dt$$

where $A(g) = \frac{3}{2}\cdot g^{\frac{2}{3}}$ in this case.
Then the Euler Equation of calculus of variations becomes

$$(3.11) \qquad \frac{d^2}{dt^2}\left(\frac{\partial}{\partial g} A(g)\right)\Big|_{g=y''} = 0,$$

i.e. $\frac{d^2}{dt^2}\left[(y'')^{-\frac{1}{3}}\right] = 0$, hence $y'' = (c - dx)^{-3}$, c,d real,

which is characteristic for special rational splines.

§4. Definition of Regular Splines

With the notations I and $\pi = (x_1,\dots,x_{m-1})$ introduced before we may generalize the concept of spline by replacing the rational functions by other classes τ_j of functions, say $t_j(x,c,d) \in C^k(I_j)$. Then we consider the *class of spline functions*

$$(4.1) \qquad S = S(\pi;t_1,\dots,t_m) = \{s \mid s \in C^k,\ s|_{I_j} = p_j(x) + t_j(x,c_j,d_j),\ p_j(x) \text{ polynomial of degree } k-1 \text{ for } j = 1,\dots,m\}.$$

We impose the following assumptions:

I) Regularity. The difference of the kth order derivatives of two elements of τ_j cannot have two separated zeros.

Consequence: We may replace the parameters c,d by the kth order derivative at x_{j-1} and x_j.

II) Smoothness. The functions $t_j(x,c,d) \in C^{(k+2)}(I_j)$ and $\frac{\partial^v}{\partial x^v} t_j(x,c,d)$ for $v \le k+2$ have continuous partial derivatives with respect to c,d.

Consequence: The higher order derivatives of t_j are parametrized by $t^{(k)}(x_{j-1})$ and $t^{(k)}(x_j)$, hence boundedness of $t^{(k)}$ implies boundedness of $t^{(k+1)}$ and $t^{(k+2)}$.

The interpolation problem P_1 carries over immediately to $S(\pi,t)$ if instead of the rational splines these more general splines (with $k = 2$) are used.

As remarked in Section 2 an inmmediate consequence of regularity is unicity of any solution to the interpolation problem P_1, i.e., Theorem 2.2 carries over without change.

The necessary conditions will depend upon the range of the parameters of $t''(x,c,d)$ and should be formulated appropriately in each case.

The constructive part of the solution to the interpolation problem very closely follows §3. We again use the 2nd order derivatives t'' at x_{j-1} and x_j to parametrize the elements of the spline s in I_j. Then $\Delta^2(x_{j-1},x_j,x_j)s$ and $\Delta^2(x_j,x_j,x_{j+1})s$ can be expressed by means of $t''_{j\pm1}$ and t''_j. Thus equations (1.4) are again applicable to the determination of the second order derivatives at the knots. It becomes, however, more difficult to make statements about the solvability of these equations.

We will describe two approaches, one by Schaback, a second one by Werner. In both cases we can rely on

Theorem 4.1: Suppose the difference quotients in the equations (1.4) are expressed as functions of the second derivatives t''_j, and let t''_j $(j = 0,\dots,m)$ form a solution of this system that has admissible components, i.e., there exist members of the classes of functions $t_j(x,t''_{j-1},t''_j) \in C^4(I_j)$. Then the interpolation problem P_1 possesses a unique solution $s(x)$. In each I_j it is given by

$$(4.2)\qquad s(x) = t_j(x,t''_{j-1},t''_j) + \frac{x - x_{j-1}}{h_j}[s_j - t_j(x_j,t''_{j-1},t''_j)]$$

$$- \frac{x - x_j}{h_j}[s_{j-1} - t_j(x_{j-1},t''_{j-1},t''_j)].$$

Proof: Uniqueness follows from regularity as remarked before. Insertion of $x = x_i$ $(i = j, j - 1)$ in (4.2) shows that the interpolation condition is met. This implies continuity of the function $s(x)$. Continuity of s'' follows from $s''(x_j) = t''_j(x_j,t''_{j-1},t''_j) = t''_{j+1}(x_j,t''_j,t''_{j+1})$. The derivative s' is given by

$$(4.3)\qquad s'(x) = t'_j(x,t''_{j-1},t''_j) + \frac{s_j - s_{j-1}}{h_j}$$

$$- \Delta^1_\tau(x_j,x_{j-1})t_j(\tau,t''_{j-1},t''_j).$$

Thus, because of the interpolation condition,

$$s'(x_k) = h\cdot\Delta^2(x_k,x_j,x_{j-1})t(.,t''_{j-1},t''_j) + \Delta^1(x_{j-1},x_j)f$$

where $h = \begin{cases} h_j & \text{if } k = j, \\ -h_j & \text{if } k = j - 1, \end{cases}$

its continuity across the knots is ensured by equation (3.4). If $s'(x_0)$ is prescribed, using (3.4) with $\lambda_0 = 0$, $\mu_0 = 1$ we get

$$s'(x_0) = \Delta^1(x_0,x_1)s - h_1\Delta^2(x_0,x_0,x_1)f = f'_0,$$

i.e., the given value. If $s''_0 = f''_0$ is prescribed, it is entered into equation (1.4) as t''_0. The same conclusions hold for $x = x_m$. □

§5. Generating Functions

One way to generalize the rational splines suggested by Schaback (1973) is the following:

Take a function $g(x)$, with properties specified below, and define

$$(5.1) \qquad t(x,c,d) = g(x/c + d) \quad \text{or}$$

$$(5.2) \qquad t(x,c,d) = c\cdot g(x + d).$$

Both classes may be used to define t_j as in the previous section. To ensure *smoothness* let g be defined in $D \subset \mathbb{R}$ and $g \in C^4(D)$. Usually $D = \mathbb{R}$ or $\mathbb{R}_+$. Of course x,c,d have to be chosen so that $x/c + d \in D$. Furthermore let $t(x,c,d)$ be *stiff*, i.e., g'' be monotone in D. Assume the *regularity* condition is met. Denote by $\varphi(y)$ the inverse function of g'', defined on $g''(D)$.

As before we like to express the 2nd order divided difference quotients of the spline s generated by (5.1) in I_j through t''_{j-1}, t''_j. The length h_j of I_j and its location should not enter. We show that this imposes one further condition on $g(x)$. For any two values $y_1, y_2 \in g''(D)$, $y_1 \neq y_2$ consider

$$y_1 = g''(x_1/c + d)/c^2 \quad \text{and} \quad y_2 = g''(x_2/c + d)/c^2.$$

Because of $x_i/c + d = \varphi(c^2\cdot y_i)$ we have

$$(5.3) \qquad \Delta^2(x_1,x_1,x_2)t = \left[\frac{t(x_2) - t(x_1)}{x_2 - x_1} - t'(x_1)\right](x_2 - x_1)^{-1}$$

$$= \left[\frac{g(\varphi(c^2\cdot y_2)) - g(\varphi(c^2\cdot y_1))}{c\varphi(c^2\cdot y_2) - c\varphi(c^2\cdot y_1)} - \right.$$

$$- \frac{g'(\varphi(c^2 \cdot y_1))}{c}\Bigg] (c\varphi(c^2 \cdot y_2) - c\varphi(c^2 \cdot y_1))^{-1},$$

and this should be a function of y_1, y_2 only, i.e., independent of c for all admissible values of c, in particular for $c = 1$. Hence we ask for the identity

$$(5.4) \qquad \left[\frac{g(\varphi(c^2 y_2)) - g(\varphi(c^2 y_1))}{\varphi(c^2 y_2) - \varphi(c^2 y_1)} - g'(\varphi(c^2 y_1))\right]$$

$$(\varphi(c^2 y_2) - \varphi(c^2 y_1))^{-1} = c^2 \left[\frac{g(\varphi(y_2)) - g(\varphi(y_1))}{\varphi(y_2) - \varphi(y_1)}\right.$$

$$\left. - g'(\varphi(y_1))\right] (\varphi(y_2) - \varphi(y_1))^{-1}$$

to hold.

Theorem 5.1: Let $g(x)$ satisfy the above conditions for smoothness, regularity and stiffness, furthermore assume equation (5.4) is valid for all values of admissible c, y_1, y_2. Then the existence of an interpolating spline for the classes given by $g(x)$ via (5.1) is equivalent to existence of a stationary point of

$$(5.5) \qquad E(M) = \sum_{i=1}^{m} h_i \frac{g(z_i) - g(z_{i-1})}{z_i - z_{i-1}} - \sum_{i=0}^{m} A_i z_i,$$

where $M = (M_0, \ldots, M_m)$, $z_i = \varphi(M_i)$ and $A_i = (h_i + h_{i+1}) \cdot \Delta^2(z_{i-1}, z_i, z_{i+1}) f$, $h_{m+1} = h_0 = 0$, $z_{-1} = z_0$, $z_{m+1} = z_m$. For $z_i = z_{i-1}$ the difference quotient should be replaced by the derivative at z_i.

Proof: Clearly,

$$\frac{\partial E}{\partial M_j} = \frac{\partial z_j}{\partial M_j} \left[h_{j+1} \left(\frac{g(z_{j+1}) - g(z_j)}{(z_{j+1} - z_j)^2} - \frac{g'(z_j)}{z_{j+1} - z_j} \right) \right.$$

$$\left. + h_j \left(\frac{g'(z_j)}{z_j - z_{j-1}} - \frac{g(z_j) - g(z_{j-1})}{(z_j - z_{j-1})^2} \right) - A_j \right].$$

Hence the determining equations (3.4) are equivalent to $\frac{\partial E}{\partial M_j} = 0 \; \forall j$. □

We can see that $E(z)$, written with this parametrization, is convex if $g'(x)$ is convex. In this case well known optimization techniques are available to solve the equations (3.4) for z_j.

Theorem 5.2: If $g'(x)$ is convex, then the function $E(z)$ defined in (5.5) and parametrized by z (not M) is convex.

Proof: To see that write

$$G(z_1, z_2) = \frac{g(z_2) - g(z_1)}{z_2 - z_1} = \int_0^1 g'(z_1 + (z_2 - z_1)t)dt \tag{5.6}$$

and let $s \in (0,1)$. If $z_1 = s \cdot a_1 + (1 - s) \cdot b_1$, $z_2 = sa_2$ $+(1 - s) \cdot b_2$ then due to convexity of g'

$$G(z_1, z_2) \le \int_0^1 s \cdot g'(a_1 + (a_2 - a_1)t)dt + \int_0^1 (1 - s)$$

$$g'(b_1 + (b_2 - b_1)t)dt \le s \cdot G(a_1, a_2) +$$

$$+ (1 - s) \cdot G(b_1, b_2).$$

Because the linear functions $- A_i z_i$ are convex and because the sum of convex functions is convex, the convexity of E is established. □

Example: Let $g(x) = \exp(x)$, $D = \mathbb{R}$

$$g''(x) = \exp(x), \; \varphi(y) = \ln y, \; g''(D) = \mathbb{R}_+.$$

Verification of (5.4): (with c^2 replaced by $c > 0$)

$$p(cy_1, cy_2) = \frac{\exp(\ln(cy_2)) - \exp(\ln(cy_1))}{(\ln(cy_2) - \ln(cy_1))^2}$$

$$- \frac{\exp(\ln(cy_1))}{\ln(cy_2) - \ln(cy_1)} = c \cdot p(y_1, y_2).$$

Since $g'''(x) > 0$ the function $g'(x)$ is convex.

Investigation of $E(z)$: Assume $A_i = (h_i + h_{i+1})$. $\Delta^2(x_{i-1},x_i,x_{i+1})f$ to be positive for all i, the number of components of z depending upon boundary data (i.e., from 0 or 1 to $m - 1$ or m) at multiplication point.

Define

$$(5.5') \qquad E(z) = \sum\left[h_i \frac{\exp(z_i) - \exp(z_{i-1})}{z_i - z_{i-1}} - A_i z_i\right].$$

(If $z_i = z_{i-1}$ use derivative instead of difference quotient.) We claim that $E(z)$ takes on its minimum at a finite point z.

Since $E(0,\ldots,0) = \sum_i h_i = H < \infty$, we restrict attention to

$$K = \{z \mid E(z) \le H\}$$

and show that this point set is bounded. Given a point $z = (z_0,\ldots,z_m) \in K$,

let $\underline{z} = \min(0,z_i)$, $\overline{z} = \max(0,z_i)$,

$$\underline{A} = \min A_i > 0,\ \underline{h} = \min_{i>0} h_i > 0,\ \sum A_i = A.$$

If $\overline{z} \neq \underline{z}$, then

$$H \ge E(z) \ge -\underline{A}\cdot\underline{z} - A\cdot\overline{z}, \text{ hence}$$

$$(5.7) \qquad \underline{z} \ge -H/\underline{A} - A\cdot\overline{z}/\underline{A}.$$

On the other hand either $\overline{z} = 0$ (hence $z_i \le 0\ \forall\ i$) or

$$H \ge E(z) \ge \underline{h}\cdot\frac{\exp(\overline{z}) - \exp(\underline{z})}{\overline{z} - \underline{z}} - A\cdot\overline{z},$$

which leads to

$$(A\cdot\overline{z} + H)(\overline{z} - \underline{z}) \ge \underline{h}\exp(\overline{z}) - \underline{h}$$

such that by (5.7) the quantity $\overline{z}$ has to satisfy

$$(A\cdot\overline{z} + H)\cdot(A\overline{z} + H + \underline{A}\overline{z})/\underline{A} + \underline{h} \ge \underline{h}\cdot\exp(\overline{z}),$$

which implies that $\overline{z}$ is bounded. By (5.7) this in turn implies boundedness of $\underline{z}$.

§6. Interpolation on Small Intervals. Convergence Properties

We will now consider partitions π of I for which

(6.1) $$|\pi| = \bar{h} = \max h_i$$

is small and ask whether there exists an interpolating spline to a given function $f(x)$ and what happens to the interpolating splines if we consider a sequence of partitions with $\bar{h} \to 0$.

For that purpose we look again at the equations (3.4) and rewrite first the difference quotients. By Taylor expansion about x for $z = x + h$ we have

(6.2) $$f(z) = f(x) + h\cdot f'(x) + \frac{h^2}{2} f''(x) + \ldots + \frac{h^V}{V!} f^V(x)$$

$$+ \frac{h^{V+1}}{V!}\cdot \int_0^1 t^V \cdot f^{(V+1)}(z - ht)dt,$$

and in particular, applied to a derivative,

(6.3) $$f^{(k)}(z) = f^{(k)}(x) + h\cdot f^{(k+1)}(x) + h^2\cdot \int_0^1 t\cdot f^{(k+2)}(z - ht)dt.$$

These formulas are readily combined to give

Lemma 6.1: Let $f(x) \in C^4(D)$ and $x,z \in D$, then

(6.4) $$\Delta^2(x,x,z)f = \frac{1}{2}f''(x) + \frac{h}{6}\cdot f'''(x) + \frac{h^2}{6}\cdot \int_0^1 t^3 f^{(4)}(z - ht)dt,$$

(6.5) $$\Delta^2(x,x,z)f = \frac{2f''(x) + f''(z)}{6} + h^2\cdot A(z,x)$$

where $h = z - x$ and

(6.6) $$A(z,x) = \frac{1}{6}\int_0^1 (t^3 - t)\cdot f^{(4)}(z - ht)dt = -\frac{1}{24}\cdot f^{(4)}(z^*)$$

with some z^* between x and z. □

Remark 6.1: The lemma is also an easy consequence of the theorem of Peano. Practically the foregoing consideration reproduces its proof for this special case.

It may be worthwhile to verify the formula (6.5) for our *standard example*. Let $r \in R_{2,1}$ and $m^3 = r''(x)$, $\underline{m}^3 = r''(z)$, then the formulae given earlier can be written as:

$$\Delta^2(x,x,z)r = \frac{1}{2}\, m^2 \cdot \underline{m} = \frac{1}{6}(2m^3 + \underline{m}^3) - \frac{1}{6}(2m + \underline{m}) \cdot (m - \underline{m})^2 .$$

Clearly $\underline{m} - m = h \cdot r'''(z^*)/(m^2 + m\underline{m} + m^2)$. Apparently the second term or the right hand side is of the order h^2 and even if differentiated w.r.t. m or $\underline{m}$ retains the order h.

Naturally, we use Lemma 6.1 to rewrite equations (3.4). We replace f by $t_j(x;x_{j-1},x_j,M_{j-1},M_j)$, thus including the endpoints of the subinterval as parameters, and denote the integral (6.6) by $A_j(x_{j-1},x_j,M_{j-1},M_j)$. This results in

$$(6.7) \qquad \lambda_j M_{j-1} + 2M_j + \mu_j \cdot M_{j+1} = 6\Delta^2(x_{j-1},x_j,x_{j+1})f$$
$$- \lambda_j h_j^2 A_j(x_{j-1},x_j,M_{j-1},M_j)$$
$$- \mu_j h_{j+1}^2 A_{j+1}(x_{j+1},x_j,M_{j+1},M_j).$$

From our standard example we learn that we cannot expect A_j to stay uniformly bounded if $x_j - x_{j-1}$ becomes small but $M_j - M_{j-1}$ does not, since this would imply the third derivatives to become infinite. On the other hand it is reasonable to assume that t_j and its derivatives, hence also A_j, are continuous in the variables M_j and $\dfrac{M_j - M_{j-1}}{h_j}$. In the following this *assumption* A is always made.

After these preparations we can formulate and prove

Theorem 6.1: Let $t(z;x_j,x_{j+1},M_j,M_{j+1})$ denote a smooth regular family of functions satisfying A. Let $f(x) \in C^4(I)$ and $f''(x)$ and $f'''(x)$ be such that

$$x, x + h, f''(x), f''(x) + hf'''(x) \text{ for } 0 \le h \le H_0,\ x \in I,$$
$$x + h \in I$$

are admissible parameters for t. Let $\gamma \in (0,1]$. Then there exists an $H > 0$ such that for every partition π with width $|\pi| \leq H$ and $h_i \geq |\pi|\cdot\gamma\ \forall i$ there exists an interpolating spline to $f(x)$. Furthermore for every sequence of such partitions with $|\pi| \to 0$ the relation

$$\|s^{(i)} - f^{(i)}\| = O(\bar{h}^{4-i}) \text{ for } i = 0,\ldots,3 \tag{6.8}$$

holds. $\|\cdot\|$ denotes the Chebyshev norm on I.

Remark 6.2: For simplicity we assumed all functions t_j to be of the same type t as in the standard example where all were rational. One could start with a coarse partition π_0 and prescribe different classes $t_j(x,c,d)$ in every subinterval of π_0. The partitions π mentioned in the theorem should then be refinements of π_0. For any subinterval of π the class used is the same as the class in π_0 for that interval which contains the subinterval of π. In this situation the conclusion of the Theorem remains true.

Proof: Denote, for one fixed partition π the vector $(M_0,M_1,\ldots,M_m)$ by M, also the vector of the values $\Delta_j^2 f = \Delta^2(x_{j-1},x_j,x_{j+1})f$ by $\Delta^2 F$, and $(f_0'',\ldots,f_m'')$ by F''. Observe that m depends on the partition π. The left hand side of the system of equations given by (6.7) is of the form $A\cdot M$ where A is a tridiagonal matrix of the form $A = 2\cdot E + C$ with row sum norm $\|C\| = 1$. If z is any vector of $\mathbb{R}^{m+1}$ and $y = Az$, then

$$y = Az = 2z + Cz \text{ implies } 2\cdot\|z\| \leq \|y\| + \|C\|\cdot\|z\|,$$

thus we deduce $\|z\| \leq \|y\|$. Hence $B = A^{-1}$ exists and $\|B\| \leq 1$, independent of m. With this result we can rewrite the system (6.7) and get

$$M = 6\cdot B\cdot\Delta^2 F - B\cdot U(M) = \phi(M) \tag{6.9}$$

where the vector $U(M)$ has the components

$$\begin{aligned} U(M)_j = {} & \lambda_j h_j^2\cdot A_j(x_{j-1},x_j,M_{j-1},M_j) \\ & + \mu_j h_{j+1}^2\cdot A_{j+1}(x_{j+1},x_j,M_{j+1},M_j). \end{aligned} \tag{6.10}$$

By our smoothness assumption the terms A_j have derivatives w.r.t., the $M_j\ \forall j$, hence $U(M)_j$ also, and the magnitude of these derivatives tends to zero as h tends to zero. This suggests that by appropriate restriction of h the mapping $\phi(M)$ may become contracting.

Futhermore we hope the solution of (6.9) to be close to F'' for small h. Hence we may set up

$$(6.11) \qquad K_\delta = \{(x,y,z) \mid x \in I,\ y \in [f''(x) - \delta, f''(x) + \delta],\ z \in [f'''(x) - \delta, f'''(x) + \delta]\}$$

and assume that δ be chosen such that for every positive $h \le H_0$, the quadruple

$$(x, x + h, y, y + hz) \text{ for every } (x,y,z) \in K_\delta \text{ and } x + h \in I$$

be an admissible set of parameters for the function t. Furthermore we may assume that uniformly for these parameters the functions t and $U(M)_j$ and their derivatives, in particular with respect to M_j and M_{j-1} be uniformly bounded by a constant $C/3$.

Let K denote the set of vectors M such that $(x_{j-1}, x_j, M_{j-1}, M_j)$ satisfy

$$(6.12) \qquad |x_{j-1} - x_j| \le H_0, \text{ and } x_{j-1}, M_{j-1}, \frac{M_{j-1} - M_j}{x_{j-1} - x_j} \in K_\delta,\ x_j \in I,$$

for $j = 1, \ldots, m(\pi)$.

In addition we may assume H_0 to be so small that $k = H_0 \cdot C < 1$. Therefore for every pair M, M^* of elements of K we have

$$(6.13) \qquad |\phi(M) - \phi(M^*)| \le h \cdot C\ |M - M^*| \le k \cdot |M - M^*|;$$

here $|\cdot|$ is the ∞-norm of $\mathbb{R}^{m+1}$.

We would like to solve (6.9) by iteration

$$(6.14) \qquad M^{j+1} = \phi(M^j),$$

starting with $M^0 = F''$. This necessitates possibly still more decreasing of H_0, as we state next.

If the iteration would always lead to an element of K then we would get the usual estimate (Werner [18])

(6.15) $$|M^j - M^0| \le \frac{|M^0 - M^1|}{1 - k} = \frac{|F'' - \phi(F'')|}{1 - k} .$$

The numerator on the right tends to zero like $O(h^2)$, $h = |\pi|$. To see this, observe that $M^1 = \phi(F'')$ satisfies

$$A \cdot M^1 = \Delta^2 F - U(F''), \text{ hence}$$

$$A \cdot (M^1 - F'') = \Delta^2 F - A \cdot F'' - U(F'').$$

By choice of K and the uniform boundedness of the derivatives by C we have $U(F'') = O(h^2)$ and from the theory of cubic splines (Werner-Schaback [19])

$$\Delta^2 F - A \cdot F'' = O(h^2) \text{ if } f \in C^4(I).$$

Therefore from (6.15) we may conclude

(6.16) $$|M^j - F''| = O(h^2)$$

if the vectors $M^0, \ldots, M^{j-1}$ were elements of K.

Obviously we may bound h by $H \le H_0$ so that

(6.17) $$|M_i^j - f_i''| < \delta \ \forall i = 0, \ldots, m \text{ and } \forall j .$$

But we also have

$$z = \frac{M_i^j - M_{i-1}^j}{x_i - x_{i-1}} = \frac{M_i^j - f_i''}{h_i} + \frac{f_i'' - f_{i-1}''}{h_i} + \frac{f_{i-1}'' - M_{i-1}''}{h_i}$$

$$= O\left(\frac{h^2}{h_i}\right) + f_{i-1}''' + O(h) + O\left(\frac{h^2}{h_i}\right),$$

and by the uniformity of π, i.e., $h_i \ge \gamma \cdot h$, we can achieve

(6.18) $$|z - f_{i-1}'''| = O(h) < \delta$$

possibly by a further reduction of H.

Having fixed the width of π in this way, one can verify that the iteration (6.14) will stay in K, i.e., each iterate M^j belongs to K and (from the contracting mapping theorem) there is a limit vector $M^* \in K$ that satisfies (6.16) and solves (6.9). Now (6.8) follows by standard arguments:

Using M^* construct the interpolating splines. Let $w(x) = s(x) - f(x)$. From (6.16) we have

$$w''_i = O(h^2) \text{ for } i = j-1, j. \tag{6.19}$$

This and $h_i \geq \gamma \cdot h$ imply the existence of $z^* \in I_j$ such that

$$w'''(z^*) = O(h). \tag{6.20}$$

Since $M^* \in K$ we know that $s^{(4)}(x)$ and $f^{(4)}(x)$ are uniformly bounded on I, the equation (6.20) implies via

$$w'''(x) = w'''(z^*) + \int_{z^*}^{x} w^{\mathrm{iv}}(t)dt$$

that $\qquad \|w'''\| = O(h)$ and similarly $\|w''\| = O(h^2)$.

Using the interpolation property of $s(x)$ one easily concludes

$$\|w\| = O(h^4) \text{ and } \|w'\| = O(h^3). \quad \square$$

This existence proof shows how one could actually compute an interpolating spline. The functions A_j used in (6.10) have to be computed for every class $t(z; x_{j-1}, x_j, M_{j-1}, M_j)$ just as we did for our standard example.

§7. Generalisation of Interpolation Problem to $k > 2$.

We will now indicate how the interpolation problem for $k > 2$ may be posed and solved. The general theory was developed in the dissertation of Arndt [1] to which we refer the reader for more details. But we will sketch the whole development and give some new proofs.

Assume a class S of regular splines as defined in (4.1) to be chosen and kept fixed in the following investigation.

Counting free parameters, one expects the following problem to be solvable.

P_2: Given values $s(x_j)$ for $j = 0, \ldots, m$ (*interpolation data*) and

$$\left.\begin{array}{l} s'(x_0), \ldots, s^{(k_0)}(x_0) \\ s'(x_m), \ldots, s^{(k_m)}(x_m) \end{array}\right\} \quad k_0 \geq 0,\ k_m \geq 0,\ k_0 + k_m = k$$

(*boundary data*),

find $s \in S$ that attains these values.

The steps in attacking this problem are completely parallel to those of the case $k = 2$.

1) Apply difference operators to eliminate the polynomial parts of the splines, in this way splitting the problem into a non-linear and linear one.

2) Expand the non-linear equations into a main part, that is linear, and a non-linear, but *contracting* part in analogy to (6.7). To do so the kth order difference quotients that we will encounter will be again expressed by means of the kth order differential quotients plus a term of order h^2. We then depend essentially on the solvability of P_2 for polynomial splines of $(k + 1)$st order; our problem takes the form of a perturbation of this linear problem.

   We make the assumption that uniformly in π for $|\pi| \leq H_0$ the linear problem possesses a solution, i.e., we assume that the arising matrix A as before has a bounded inverse B.

3) Finally we have to construct the splines, i.e., determine the polynomial parts in each subinterval from the given interpolation and boundary data and the calculated values of $s_i^{(k)}$, $i = 0,\dots,m$. (Terminology and notation as before.)

Without going into detail we remark that under the above assumptions the regularity of the splines implies *uniqueness*. Counting zeros as before, the kth order derivative of the difference w of two interpolating splines can be shown to vanish and we are left with a polynomial spline of order $k - 1$ for this difference w that has to attain zero values for all data given in P_2.

The first step is based on the generalisation of equation (1.4) by

Lemma 7.1: Every kth order difference quotient can be written as a linear combination of *confluent* kth order difference quotient each having as arguments only two adjacent knots x_{j-1}, x_j (with appropriate multiplicity); the coefficients of the combination are positive, their sum is equal to one (one may call this a *convex* linear combination.)

Remark 7.1: The proof of the lemma gives also a recursive construction of the coefficients.

Proof: Denote the set

$$\mathbb{Z} = \{z_1, z_2, \ldots, z_{km+k}\} = \{\underbrace{x_0, \ldots, x_0}_{k \text{ times}},$$

$$\underbrace{x_1, \ldots, x_1}_{k \text{ times}}, \ldots, \underbrace{x_m, \ldots, x_m}_{k \text{ times}}\}$$

and define a *chain* $\mathbb{Z}_{i,j}$ to be the subset

$$\mathbb{Z}_{i,j} = \{z_i, z_{i+1}, \ldots, z_j\}$$

with the *length* $j - i$.

The *length of an arbitrary subset* $X \subset \mathbb{Z}$ is denoted by $l(X)$ and defined by the minimal length chain $\mathbb{Z}_{i,j}$ containing X. For X we take the arguments of kth order difference quotients.

If X contains only arguments of value x_{j-1} and x_j (with appropriate multiplicity) then X is already a chain of $k + 1$ elements hence $l(X) = k$.

For this kind of difference quotient the statement of the theorem is trivial, since the sum consists of only one term.

Now assume $L = l(X) > k$. We proceed by induction and assume the lemma is correct for every difference quotient with argument X such that $l(X) < L$.

Assume x_i to be the smallest, x_j the largest component of X. Since $l(X) > k$ these two knots cannot be adjacent, hence there is x_r of π such that $x_i < x_r < x_j$. Let $X' = X \backslash \{x_i, x_j\}$ i.e., the set from which x_i and x_j are removed.

Then we have the identity

$$\Delta^k(X)f = \Delta_t(x_i, x_j)\Delta^{k-1}(t, x')f \tag{7.2}$$

$$= \left[\frac{x_j - x_r}{x_j - x_i}\,\Delta_t(x_r, x_j) + \frac{x_r - x_i}{x_j - x_i}\,\Delta_t(x_i, x_r)\right].$$

$$\Delta^{k-1}(t, X')f$$

$$= \frac{x_j - x_r}{x_j - x_i} \Delta^k(x_r, X', x_j)f + \frac{x_r - x_i}{x_j - x_i} \Delta^k(x_i, X', x_r)f$$

and the length of (x_r, X', x_j) respectively to (x_i, X', x_r) is smaller than $l(X)$. Since a smallest, respectively a largest element of X has been replaced by an intermediate point was already contained in the minimal chain comprising X. Hence both kth order difference quotients of the right hand side can be expressed as convex linear combinations of confluent difference quotients. It is clear that a convex linear combination of convex linear combinations gives a convex linear combination. Hence $\Delta^k(X)f$ can be expressed as stated by the lemma. □

As we know the multiplicity r of an argument x_j of a difference quotient indicates that $f_j, f'_j, \ldots, f_j^{(r-1)}$ are used for the calculation. If we symbolize the difference quotient by a polygonal curve which connects the data used, then we may indicate the recursion generated by the above proof for one example by the following diagram. (We dispense with the easy task of keeping track of the coefficients that are generated.)

Example: Consider $\Delta^3(x_1,x_2,x_3,x_4)f = \Delta^3(X)f$.

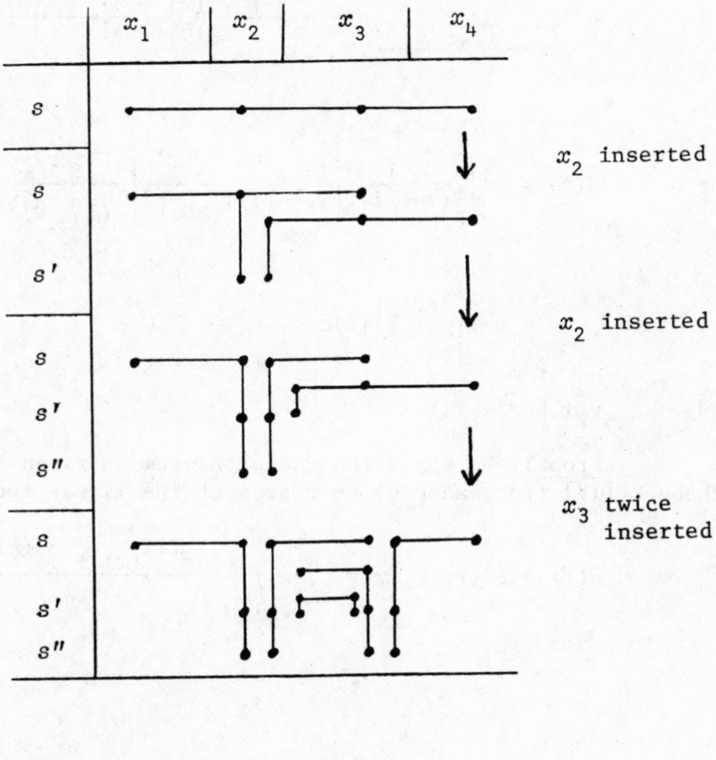

$\Delta^3(X)f$ is expressible as linear combination of difference quotients with arguments

$$
\begin{array}{l}
x_1, x_2, x_2, x_2 \\
\quad x_2, x_2, x_2, x_3 \\
\qquad x_2, x_2, x_3, x_3 \\
\qquad\quad x_2, x_3, x_3, x_3 \\
\qquad\qquad x_3, x_3, x_3, x_4 .
\end{array}
$$

The identities described by Lemma 7.1 may be applied to the data given in P_2 on the one hand and to the splines given in (4.1) on the other hand. The result is a system of equations similar to (3.4). The terms on the left hand side are now kth order difference quotients containing the searched parameters $t_i^{(k)}$ in a possibly highly non-linear way.

We proceed to the second step by proving

Lemma 7.2: Let $f \in C^{k+2}(I)$; $x, z \in I$, write $h = z - x$ (assumed positive). Then

$$\Delta^k(\underbrace{x,\ldots,x}_{\mu},\underbrace{z,\ldots,z}_{\nu})f = \frac{\mu \cdot f^{(k)}(x) + \nu \cdot f^{(k)}(z)}{(k+1)!} + R(f),$$

where $$R(f) = h^2 \int_0^1 \frac{1}{(k+1)!}\left[\frac{(-1)^{\nu-1}}{(\nu-1)!}\,\frac{\partial^{\nu-1}}{\partial v^{\nu-1}}\left(\frac{(t-v)_+^{k+1}}{(1-v)^{\mu}}\right)\bigg|_{v=0} - \nu \cdot t\right] f^{(k+2)}(t)\,dt$$

and $\nu + \mu = k + 1$.

Proof: We apply the Peano theorem as given in Werner-Schaback [19] (for example) to represent the linear functional

$$R(f) = \Delta^k(x,\ldots,x,z,\ldots z)f - \frac{\mu f^{(k)}(x) + \nu f^{(k)}(z)}{(k+1)!}$$

$$= \sum_{r=0}^{k+1} a_r f^r(x) + \int_x^z R_u\left\{\frac{(u - t)_+^{k+1}}{(k + 1)!}\right\} f^{(k+2)}(t)dt.$$

Since $a_r = R_u\left(\frac{(u - x)^r}{r!}\right)$, we see that

$$a_r = 0 \quad \text{for } r = 0,\ldots,k - 1.$$

It is also clear that we have

$$R_u\left(\frac{(u-x)^k}{k!}\right) = \Delta^k(\ldots)(u - x)^k - \frac{\mu \cdot k! + \nu k!}{(k + 1)!} = 1 - 1 = 0.$$

For the next degree we write, for $f = \frac{(t - x)^{k+1}}{(k + 1)!}$,

$$\Delta^k(x,\ldots,x,z,\ldots,z)f = \Delta_\tau^{\nu-1}(z,\ldots,z)\cdot\Delta^\mu(\tau,x\ldots,x)f$$

$$= \frac{1}{(\nu - 1)!}\cdot\frac{\partial^{\nu-1}}{\partial\tau^{\nu-1}}\left(\frac{1}{(\tau - x)^\mu}\cdot\frac{(\tau - x)^{k+1}}{(k + 1)!}\right)\Bigg|_{\tau=z}$$

$$= \frac{1}{(k + 1)!}\cdot\frac{\partial^{\nu-1}}{\partial\tau^{\nu-1}}\frac{(\tau - x)^\nu}{(\nu - 1)!}\Bigg|_{\tau=z} = \frac{\nu}{(k + 1)!}\, h.$$

Therefore

$$R_u\left(\frac{(u - x)^{k+1}}{(k + 1)!}\right) = \frac{\nu!}{(k + 1)!}\,\frac{h}{(\nu - 1)!} - \frac{\nu}{(k + 1)!}\,\frac{\partial^k}{\partial u^k}$$

$$\frac{(u - x)^{k+1}}{(k + 1)!}\Bigg|_{u=z} - \frac{\mu}{(k + 1)!}\cdot 0 = 0.$$

The kernel function is given by

$$R_u\left(\frac{(u - \tau)_+^{k+1}}{(k + 1)!}\right) = \frac{1}{(k + 1)!}\left[\frac{1}{(\nu - 1)!}\,\frac{\partial^{\nu-1}}{\partial u^{\nu-1}}\right.$$

$$\left.\left(\frac{(u - \tau)_+^{k+1}}{(u - x)^\mu}\right)\Bigg|_{u=z} - \nu\cdot(z - \tau)\right] \quad \text{for } x < \tau < z.$$

We substitute

$$z - \tau = h\cdot t \qquad \text{where} \quad \begin{cases} t = 0 \sim \tau = z \\ t = 1 \sim \tau = x \end{cases}$$
$$z - u = h\cdot v$$

to find

$$R_u(\ldots) = \frac{h}{(k+1)!}\left[\frac{(-1)^{\nu-1}}{(\nu-1)!}\,\frac{\partial^{\nu-1}}{\partial v^{\nu-1}}\left(\frac{(t-v)_+^{k+1}}{(1-v)^{\mu}}\right)\bigg|_{v=0} - \nu\cdot t\right].$$

This yields the above formula if one observes that the order of integration had to be reversed and $d\tau = -hdt$ holds. □

<u>Examples</u>:

1. $$\Delta^3(x,x,x,z)f = \frac{3f'''(x) + f'''(z)}{24}.$$

$$+ \frac{h^2}{24}\int_0^1 (t^4 - t)f^{(5)}(z - ht)dt .$$

Since $\mu = 3$, $\nu = 1$, $k + 1 = 4$,

$$R_u(\ldots) = \frac{h}{24}\left[\left(\frac{(t-v)_+^4}{(1-v)^3}\right)\bigg|_{v=0} - t\right] = \frac{h}{24}\cdot(t^4 - t).$$

2. $$\Delta^3(x,x,z,z)f = \frac{f'''(x) + f'''(z)}{12}$$

$$+ \frac{h^2}{12}\int_0^1 (2t^3 - t^4 - t)f^{(5)}(z - ht)dt.$$

Here $\mu = \nu = 2$, $k + 1 = 4$ and

$$R_u(\ldots) = \frac{h}{24}\left[\frac{-1}{1!}\,\frac{\partial}{\partial v}\left(\frac{(t-v)_+^4}{(1-v)^2}\right)\bigg|_{v=0} - 2t\right]$$

$$= \frac{h}{24}\left[-2t^4 + 4t^3 - 2t\right].$$

Observe the symmetry of the kernel

$$\phi(t) = -t^4 + 2t^3 - t = -t(1-t)\cdot[t(1-t) + 1]$$

$$= \phi(1 - t)$$

which has to hold, of course.

It is left to the reader to use the formulas given before to formulate the defining equations for the derivatives $s_j^{(k)}$ $(j = 0,\ldots,m)$. On the left hand side the equations take the form $A \cdot M$ where A is the matrix that would result if $(k + 1)$st order polynomial splines were used to solve P_2. For small intervals the previous ideas can be carried over to prove the existence of a solution, (§6).

We come to the third step -- the construction of the linear parts of the interpolating spline, where now in addition to the interpolation and boundary data all k-th order derivatives at the knots are known.

Let k_0 derivatives $s'(x_0),\ldots,s^{k_0}(x_0)$ be given as boundary values at x_0. Then $\Delta^k(\underbrace{x_0,\ldots,x_0}_{k_0+1},\underbrace{x_1,\ldots,x_m,x_m,\ldots}_{k-k_0})f$ is determined by the given interpolation and boundary data.

The difference quotients

$$\Delta^k(\underbrace{x_0,\ldots,x_0}_{k_0+2},x_1,\ldots)s,\Delta^k(\underbrace{x_0,\ldots,x_0}_{k_0+3},x_1,\ldots)s,\ldots$$

are known via the k-th order derivatives and since in the first quotient every value of s and its derivatives except $s^{(k_0+1)}(x_0)$ are known, we can determine $s^{(k_0+1)}(x_0)$. From the next difference quotient we can then obtain $s^{(k_0+2)}(x_0)$, and so forth. Hence we get enough data to determine $s(x)$ in I_1. By continuity $s|_{I_1}$ will determine the values of all derivatives at x_1 needed to determine s in $I_2,\ldots$. So, at least theoretically, this shows how this linear problem may be solved and $s(x)$ be found.

We confine ourselves to this remark and refer the interested reader again to the dissertation of Arndt.

§8. Remarks on Chebyshev-Approximation by Regular Splines: Axiom of Steepness and Stiffness

The following sections will be devoted to some comments on approximation of continuous functions $f \in C(I)$ by regular splines. Typically there are the questions of existence, uniqueness and characterization.

The good convergence properties found for interpolation by regular splines may foster the expectation that Chebyshev approximation from this class may have favorable properties also.

Since uniqueness does not even persist in the special case of polynomial splines unless strong assumptions are made we will not pursue this question further.

Existence proofs are usually based on compactness arguments and it is this reason that leads us to consider the closure of classes of regular splines under compact convergence.

Convergence of polynomials of degree at most n on any set of $n + 1$ distinct points will lead to uniform convergence on any finite real interval and analogous conclusions hold for Chebyshev systems. The uniform convergence carries over even to the derivatives, if they exist. These facts are also typical for linear splines with fixed knots. The situation is quite different for rational functions and rational splines. We study our standard example to see what cases there are.

If we transform the fractional term by adding a specified linear polynomial, the class of rational splines is unchanged. In other words, we only change the representations of the class. For our purpose it is convenient to write

$$(8.1) \qquad t(z,c,d) = \frac{c \cdot z^2}{d - z} = \frac{c \cdot d^2}{d - z} - c(d + z) \text{ on } z < 0 \text{ for}$$

$$c > 0,\ d > 0, \text{ say, because}$$

$$t(0,c,d) = t'(0,c,d) = 0 \ \forall\ c,d.$$

This is a kind of a *normalization*.

For simplicity consider the one parameter family where $c = 1$. One easily verifies the following limits to hold for $d \to 0$:

(8.2)

	$z < 0$	$z = 0$
$\lim t\ (z,1,d)$	$-z$	0
$\lim t'(z,1,d)$	-1	0
$\lim t''(z,1,d)$	0	∞

Evidently, a similar construction is possible for other classes, i.e., exponential expressions $t(z,c,d) = c(e^{z/d} - 1 - z/d)$ on $z < 0$ if $d > 0$ and c is coupled to d in an appropriate way; here $c = 1$.

We observe that the highest derivative diverges in one end point, and tends to zero at every other point of the subinterval in such a way that the derivatives of lower order exist and are continuous except for a possible jump at said end point.

We try to formalize this situation by an axiom:

Definition: The family $t(x,\tilde{c},\tilde{d})$ is *steep* at the boundary if one can find a polynomial $p(x,\tilde{d})$ of degree $k = 1$ (where coefficients may depend on $\tilde{d}$) and a new parametrization $\tilde{c}(c,d)$, $\tilde{d}(c,d)$ such that $t(x;x_j,c,d) = t(x,\tilde{c},\tilde{d}) + p(x,\tilde{d}) = c\ T(z;x_j,d)$, continuous in x_j,d, $z = x - x_j$, $c > 0$, $d \in (0,\infty)$ and if $d \to 0$ implies

(8.3)

	for $x \in [x_{j-1},x_j)$	$x = x_j$	
$\lim T^{(v)}(z;x_j,d)$	$\frac{-z^{k-1-v}}{(k-1-v)!}$	0	for $v = 0,\ldots,k-1$
$\lim T^{(k)}(z;x_j,d)$	0	∞	

and similarly with $z = x - x_{j-1}$ for $d \to \infty$

(8.4)

	$x = x_{j-1}$	$x \in (x_{j-1},x_j]$	
$\lim T^{(v)}(z;x_{j-1},d)$	0	$\frac{z^{k-1-v}}{(k-1-v)!}$	$v = 0,\ldots,k-1$
$\lim T^{(k)}(z;x_{j-1},d)$	∞	0	

The conditions for the kth and the $(k - 1)$th order derivatives provide for a jump of these derivatives at the knots. The conditions on the lower order derivatives provide a kind of normalization which proves useful in the subsequent proofs. As we saw above there are no restrictions on the generality of the class considered.

It is also convenient to assume in the following paragraph that the splines are *stiff*, i.e., their kth order derivatives vanish identically or not at all.

§9. The Closure of Regular Stiff Steep Splines for Fixed Knots

Given a fixed partition π and classes $c \cdot T_j(z, x_j, d)$ of regular stiff splines in each I_j, which also satisfy the axiom of steepness. Define the class S of splines by (4.1).

Consider a sequence $s_i(x) \in S$, assume boundedness of the Chebyshev norm $\|s_i\| < K$.

Take $\delta < \dfrac{\beta - \alpha}{2k}$, then the following estimates hold for each member of the sequence:

$$(9.1) \qquad |\Delta^v(\alpha, \alpha + \delta, \ldots, \alpha + v\delta) s_i| < \frac{2^v \cdot K \cdot \delta^{-v}}{v!} = K_v / v!$$

for $v = 0, \ldots, k$.

This guarantees existence of $x^*_{iv} \in (\alpha, \alpha + v\delta)$ such that

$$(9.2) \qquad |s_i^{(v)}(x^*_{iv})| \le K_v .$$

With the same argumentation we get a point y^*_{iv} close to β for which $|s_i^{(k-1)}(y^*_{i,k-1})| \le K_{k-1}$ holds.

Due to stiffness $s_i^{(k)}(x)$ does not change sign. Hence the foregoing estimates and (9.2) imply uniform bounds for

$$s_i^{(k-1)}(x) \text{ for } x \in [\alpha + v\delta, \beta - v\delta] \text{ and } \forall i \in \mathbb{N}.$$

This in turn implies existence of a subsequence (which we imply without renumbering) such that in said interval $s_i^{(v)}$ converge uniformly for $v = 0, \ldots, k - 2$. Using a sequence $\delta \to 0$

and the diagonal method, we may assume the convergence to hold on (α,β), uniformly on every closed subinterval. (*Compact convergence.*)

Referring to another selection (without renumbering) we assume $s_k^{(k-1)}$ and $s_i^{(k)}$ to converge to a limit at each knot x_j, the limit value ∞ being possible. Due to the one-to-one correspondence between the parameters and the kth order derivatives at the knots this implies convergence of the parameters as well.

What properties will the limit functions have? Consider a subinterval I_j. Suppose x_j is an interior knot, (if $j = m$, take x_{m-1} instead of x_m), let $z = x - x_j$ and s denote any element of the sequence (dropping the index i). Since

$$s^{(v)} = p^{(v)}(z) + c\cdot T^{(v)}(z,x_j,d) \text{ for } x \in I_j,\ v = 0,\dots,k$$

(omitting indices of T)

and $\lim T^{(v)}(0,x_j,d) = 0$ for $v < k$,

the above convergence assumptions imply uniform convergence of the polynomial parts of s on I_j.

Denote $M_j^* = \lim s^{(k)}(x_j)$.

We distinguish two cases.

1) $M_j^* = \infty$.

This means $\lim c\cdot T^{(k)}(0,x_j,d) = \infty$. Hence either c or $T^{(k)}$ tend to infinity. Boundedness of $s^{(k-1)}$ and convergence of $p^{(k-1)}$ at every interior point $x^* \in I_j$ imply boundedness of

$$c\cdot T^{(k-1)}(x^* - x_j,x_j,d).$$

If $d^* = \lim d$ is contained in $(0,\infty)$ there is $\tilde{x}$ such that $T^{(k-1)}(\tilde{x} - x_j,x_j,d^*) \neq 0$. Hence c is bounded and also $T^k(0,x_j,d^*)$ which rules out this case.

The case $d^* = 0$ or ∞ leads to $T^{(k-1)}(\tilde{x} - x_j, x_j, d^*) = \pm 1$ which implies boundedness of c. Hence $M_j^* = \infty$ implies $d^* = 0$ by (8.3).

We conclude that the limit function is reduced to a polynomial of degree $k - 1$ in the interior of I_j. If $\lim c = c_j^*$, then $s^* = P^*(x) - c_j^* \cdot z^{k-1}/(k - 1)!$ on I_j. To the right of x_j by the same arguments the functions $s(x)$ also tend to a $(k - 1)$st degree polynomial. The coefficients c may tend to a limit c_{j+1}^*. At the knot x_j the $(k - 1)$st order derivative of the limit function s^* then will have a jump of the magnitude $c_j^* + c_{j+1}^*$. The sum $c_j^* + c_{j+1}^*$ could be zero but not negative. $s^{*(k-1)}(x)$ will be a monotone function on I just as the $(k - 1)$st order derivatives of elements of S are.

We say, s^* *is of* $(k - 1)$*st kind at* x_j to indicate that $D^v s^*$ for $v < k - 1$ is continuous while $D^{k-1} s^*$ may have a jump discontinuity.

If we now consider the second endpoint of I_j we get $M_{j-1}^* = 0$ by the boundedness of the sequence of the c and (8.3). The consequences of this fact are included in the following case.

ii) $M_j^* = \lim s^{(k)}(x_j) \in [0,\infty)$.

If c^* and d^* both are positive and finite then $c^* \cdot T(z;x_j,d^*)$ belongs to S. The limit functions behave "regularly" in I_j. If $d^* = \infty$, the above arguments show that c^* is finite. If finally $d^* = \lim d = 0$, then

$$\lim T^{(k)}(0,x_j,d) = \infty,$$

hence $c^* = \lim c = 0$, and $s^*|_{I_j}$ reduces again to a polynomial of degree $k - 1$ in I_j. If $s^*|_{I_{j+1}}$ is also a polynomial of degree $k - 1$, which is certainly the case if $M_j = 0$, then $s^{(v)}(x)$ for $v < k$ is continuous at x_j and the two polynomials to the right and the left of x_j are equal.

If s^* is a polynomial on one side of x_j and regular at the other, then M_j^* must be finite and the kth order derivative has a jump while the $(k - 1)$st one is continuous at x_j. Hence s^* is of the kth kind in this case.

We thus get a complete survey of all possibilities:

Theorem 9.1: If the knots of π are fixed, the limit function of any compactly convergent sequence of S has the following properties. Its restriction to any subinterval* is either a polynomial of degree $k - 1$ or regular, i.e., of the form $p(x) + c\,T(z,x_j,d)$, $c \neq 0$.

At the knots the limit function is either of

$(k + 1)$st kind and regular or a polynomial on both sides,

kth kind and regular on one, a polynomial on the other side,
$(k - 1)$st kind and a polynomial on both sides

of the knot. Furthermore, adjacent to an interior knot of kth or $(k - 1)$st kind is a knot where s is of $(k + 1)$st kind and polynomial at both sides. The limit function satisfies $s^{(k)}(x) \geq 0$.

We will not show here that indeed every function as described by Theorem 9.1 is in the closure of S. Examples verify that continuous functions exist that have Chebyshev approximations whose kth and $(k - 1)$st order derivatives have the described discontinuities (Braess-Werner [3]). The situation becomes even worse, if we allow free movement of (a fixed number of) the knots (Werner-Loeb [17]).

All this just indicates that one should be cautious in using regular splines for Chebyshev approximation. The differentiability properties might get lost.

§10. An Application of Non-linear Splines to the Solution of Initial Value Problems of Ordinary Differential Equations

Today there are very efficient methods and even program packages for the solution of initial-value problems. A good algorithm would not overrun a singularity but would usually decrease the step size and finally come to a stop. We are here proposing a method for handling those solutions that become singular. If the singularity can be located beforehand, i.e., from the differential equation, it is not too difficult to take precautions. Non-linear differential equations may, however, lead to solutions that tend to infinity at some point although the differential equation is smooth or even analytic everywhere. In such a case one may have some information about the type of the solution and we may take advantage of this knowledge in constructing our splines.

---

*Between adjoint knots.

Most methods are based on "linear concepts" i.e., some Taylor expansion, fitting of polynomials etc., but there are a few exceptions. Lambert and Shaw [5] used rational and exponential trial functions in our context, but did not put these pieces together to form one smooth curve. They obtained, as do almost all other methods, values of the solution on certain grid-points (*"discrete methods"*). The interpolation between these values to evaluate the solution at intermediate points then needs a second thought to take advantage of the high accuracy gained at the grid-points.

One may try to fit the solution by a smooth function throughout its domain of definition and could call this a *"continuous method"*. Those functions could stem from an iteration scheme, but practically those straight-forward methods seem of no significance.

The approach presented here approximates the solution by a spline. In the special case of cubic splines it was already designed by Loscalzo and Talbot [6]; the regular splines were introduced in the dissertation of Runge [8]; the results were improved by Werner [16].

Consider the classical initial value problem in its simplest form

P_3: Given $f(x,y)$, continuous for $x \in I = [x_0, x_+]$, $y \in \mathbb{R}$, furthermore an initial value $y_0 \in \mathbb{R}$.

Find $\delta > 0$ and a continuously differentiable function $y(x)$ defined in $x_0 \leq x \leq x_0 + \delta$ which satisfies

$y(x_0) = y_0$ and

(10.1) $y'(x) = f(x,y(x))$ for $\forall\ x \in I$.

We will require the *stronger assumption* to hold that $f(x,y)$ be of class $C^{k+2}(I \times \mathbb{R})$ to ensure that $y(x)$ is $k + 3$ times continuously differentiable.

Equation (10.1) determines the value

$$y'(x_0) = f(x_0,y_0);$$

differentiation of (10.1) will recursively determine the higher order derivatives of $y(x)$ at the point x_0

$$(10.2)\qquad \frac{d^{\nu}}{dx^{\nu}}\, y(x)\Big|_{x=x_0} = \frac{d^{\nu-1}}{dx^{\nu-1}}\, f(x,y(x))\Big|_{x=x_0} .$$

Theoretically this is obvious, from the practical stand point it is not very pleasant that we have to differentiate $f(x,y)$, but it is used only at $x = x_0$, because we set up the following

Algorithmic procedure: Fix a step size h and let

$$x_j = x_0 + jh.$$

Consider a spline $u(x) = u(z;x_j,u_j,u_j',\ldots,u_j^{(k)},u_j^{(k+1)})$ of class C^k, depending upon the $k + 2$ parameters $u_j,\ldots,u_j^{(k+1)}$ with

$$z = x - x_j \text{ and } \frac{d^{\nu}}{dz^{\nu}}\, u(z;x_j,u_j,\ldots,u_j^{(k+1)})\Big|_{z=0} = u_j^{(\nu)} .$$

Determine these parameters such that

$$(10.3)\qquad \frac{d^{\nu}}{dx^{\nu}}\, u(x_j) = u^{(\nu)}(x_j - 0) \text{ for } \nu = 0,\ldots,k \text{ and}$$

$$(10.4)\qquad u'(x_{j+1}) = f(x_{j+1},u(x_{j+1}))$$

to define $u(x)$ in I_{j+1}.

For $j = 0$ the right hand side of (10.3) should be replaced by the values $y^{(\nu)}(x_0)$ known by the process described above. For $j > 0$ the values $u^{(\nu)}(x_j - 0)$ are brought forward by the approximate solution $u(x)$ calculated for I_j.

There are the questions about the solvability of the equations (10.4), the precision of the obtained solution and the numerical stability of the algorithm.

In this paragraph we restrict attention to the case $k = 2$ since for higher values of k this simple approach will lead to numerical instability and more sophisticated measures are necessary to ensure satisfactory behavior of the method.

Let us find out about typical behavior of solutions of typical differential equations.

The functions

$$y = [c(x - x_s)]^{\alpha}, \; x_s \text{ a parameter, } \alpha, c \text{ fixed, } \alpha \neq 0$$

are solutions of the differential equation

$$y' = c \cdot \alpha \cdot y^{\frac{\alpha-1}{\alpha}}$$

This leads us to expect that a spline containing terms

$$\text{const.}(x - x_s)^{\alpha}, \; \alpha = \frac{1}{1 - n}$$

should be a good trial function for the solution of the differential equation

$$y' = p(x,y)$$

where $p(x,y)$ is a polynomial in x and y with degree n in y. In the case of Riccati-equations $p(x,y)$ is quadratic in y and $\alpha = -1$, i.e., the solutions in general have poles of the first order. As mentioned before, it cannot be read off the differential equation but depends on the initial value also where a solution blows up. The solutions have *movable singularities*. It should be mentioned that the exponent $\alpha = \dfrac{1}{1 - n}$ may no longer be the best guess if the coefficient of y^n of $p(x,y)$ has a zero at the place x_s of the calculated solution.

Let us give more details by setting up two classes of splines using functions of 4 parameters in each subinterval. (corresponding to $k = 2$) and which are adapted to the above mentioned behavior of solutions.

Case 1: If order α_j of singularity is a free parameter and B_j a second one, define

$$(10.5) \qquad u(x) = \begin{cases} u_j + \dfrac{u_j' \cdot B_j}{\alpha_j}\left[\left(1 + \dfrac{z}{B_j}\right)^{\alpha_j} - 1\right] & \text{for } \alpha_j \neq 0, \\[2ex] u_j + u_j' \cdot B_j \ln\left(1 + \dfrac{z}{B_j}\right) & \text{for } \alpha_j = 0. \end{cases}$$

If $\alpha_j = 1$ results then u is linear and B_j does not influence the values of $u(x)$.

It is immediately seen that

$$u(x_j) = u_j,$$

$$u'(x_j) = u'_j,$$

$$u''(x_j) = \frac{u'_j(\alpha_j - 1)}{B_j},$$

hence we define $B_j = \frac{u'_j(\alpha_j - 1)}{u''_j}$ to incorporate into u all data given at x_j. If $u'''(x_j)$ were known also, one could even calculate α_j. (In this case it is easier to use u'''_j as a parameter and express α_j by it then to introduce u''_{j+1}, as was assumed in the foregoing theory of §6.)

But the algorithm requires α_j to be chosen such that (10.4) holds. Thus we have to satisfy the equation

$$u'(x_{j+1}) = u'_j\left(1 + \frac{h \cdot u''_j}{u'_j \beta_j}\right)^{\beta_j} \stackrel{!}{=} f(x_{j+1}, u(x_{j+1})), \tag{10.6}$$

where $u(x_{j+1})$ also contains $\beta_j = \alpha_j - 1$ via (10.5).
One may rewrite (10.6) to find

$$g(\mu) = (1 + \frac{1}{\mu})^{\mu} = \left(\frac{f(x_{j+1}, u(x_{j+1}))}{u'_j}\right)^{\frac{u'_j}{h \cdot u''_j}} = V, \tag{10.7}$$

where $\mu = \frac{\beta_j \cdot u'_j}{h \cdot u''_j}$ is used.

With the substitution $\frac{1}{\sigma} = 1 + \frac{1}{\mu}$ the function $g(\mu)$ can be expressed as

$$g(\mu) = \sigma^{\frac{\sigma}{\sigma - 1}} \quad \text{for } \sigma > 0.$$

The range of $g(\mu)$ is $(1,\infty)$.

If f were independent of μ, i.e., a numerical quadrature were performed, one could find α by inverting (10.7). It might be noted that a good guess to start the inversion of g by Newton iteration is given by the function

$$(10.8)\qquad \sigma = \psi(V) = \begin{cases} \left(\dfrac{V-1}{e-1}\right)^{\zeta}, \zeta = \dfrac{2(e-1)}{e} = 1.264\ 241\ 117\ 657 & \text{for } 1 < V \le 3.45 \\ V\left(1 - \dfrac{\ln V}{V-1.02}\right) - 0.13 & \text{for } V > 3.45. \end{cases}$$

We found convergence in no more than 3 steps for 12 decimals accuracy.

We refer to the papers by Werner - Wuytack [20] and Werner - Zwick [21] for examples of the application to numerical integration.

If f depends upon y, then solution of (10.7) will be by iteration. For $j = 0$ the value $y'''(x_0)$ could help finding a starting value of $\alpha_0^{(0)}$ to be inserted at the right hand side as was pointed out above. If $j > 0$ the value α_{j-1} used in the foregoing interval I_j will serve the same purpose.

The condition $V > 1$ for invertibility of y somewhat reflects the properties of the family of splines. Its second derivative cannot vanish. If u is any function with $u'_{j+1} = f(x_{j+1}, u_{j+1})$ then

$$(10.9)\qquad V = \left(\frac{f_{j+1}}{u'_j}\right)^{\frac{u'_j}{hu''_j}} = \left(\frac{u'_{j+1}}{u'_j}\right)^{\frac{u'_j}{hu''_j}} = \left(1 + \frac{h\cdot u''(x^*)}{u'_j}\right)^{\frac{u'_j}{hu''_j}},$$

with some $x^* \in I_{j+1}$.

Since $\dfrac{h\cdot u''_j}{u'_j}$ and $\dfrac{h\cdot u''(x^*)}{u'_j}$ have equal sign we always have $V > 1$, assuming both terms being greater than -1.

The only difficulties can arise if $y''(x)$ has a zero in I_{j+1} and this may lead to u'' having a zero too or being rather small. In this case change to the class of cubic splines could serve as a help.

Case 2: Exponent α of growth of solution determined beforehand. Then let α be fixed,

$$z = x - x_j .$$

$$u(x) = \begin{cases} u_j + u'_j \cdot z + \dfrac{b_j^2 \cdot u''_j}{\alpha(\alpha-1)}\left[\left(1 + \dfrac{z}{b_j}\right)^{\alpha} - 1 - \dfrac{\alpha}{b_j} z\right] & \text{for } \alpha \neq 0,1 \\ u_j + u'_j z - b_j^2 u''_j\left[\ln\left(1 + \dfrac{z}{b_j}\right) - \dfrac{z}{b_j}\right] & \alpha = 0 \\ u_j + u'_j z + b_j^2 u''_j\left[\left(1 + \dfrac{z}{b_j}\right)\ln\left(1 + \dfrac{z}{b_j}\right) - \dfrac{z}{b_j}\right] & \alpha = 1 \end{cases}$$

and $b_j = \dfrac{(\alpha - 2)\cdot u''_j}{u'''_j}$ furnishes the connection to the third derivative if it were known.

One should observe, that $\alpha = 2$ corresponds to u being a quadratic polynomial in which case only 3 parameters are at our disposal so that in general we cannot use this type of function for the algorithm.

The parameter b_j should be used to fit the equation (10.4). One obtains

$$u'(x) = \begin{cases} u'_j + \dfrac{b_j \cdot u''_j}{\alpha - 1}\left[\left(1 + \dfrac{z}{b_j}\right)^{\alpha-1} - 1\right] & \alpha \neq 1 \\ u'_j + b_j u''_j \ln\left(1 + \dfrac{z}{b_j}\right) & \alpha = 1 \end{cases}$$

and thus for $\alpha \neq 1$ the equation

$$\text{(10.11)} \qquad \frac{b_j}{(\alpha - 1)\cdot h}\left[\left(1 + \frac{h}{b_j}\right)^{\alpha-1} - 1\right] = \frac{f(x_{j+1}, u_{j+1}) - u'_j}{h \cdot u''_j},$$

provided $u''_j \neq 0$.

With the substitution $\gamma = \alpha - 1$, $\sigma = 1 + \dfrac{h}{b_j}$ the left hand side is

$$h(\sigma,\gamma) = \frac{1}{\gamma}\,\frac{\sigma^\gamma - 1}{\sigma - 1}, \qquad \sigma \neq 1,$$

$$h(1,\gamma) = 1,$$

and one checks that the range of h is given by

γ	range	
$(-\infty,0)$	$(0,\infty)$	
$(0,1)$	$(0,\frac{1}{\gamma})$	for $\sigma \in (0,\infty)$.
$(1,\infty)$	$(\frac{1}{\gamma},\infty)$	

Hence the range may be characterised by the two conditions

$$V > 0 \text{ and } \frac{1-\gamma V}{1-\gamma} > 0.$$

These conditions together with $u_j'' \neq 0$ are easily checked by a program that implements the algorithm for this class of splines. A first guess for the inversion of $h(\sigma,\gamma)$ is given by

$$\sigma = \frac{2}{\gamma - 1}\left(\frac{1-\gamma V}{1-\gamma}\right)^{\frac{2}{\gamma}-1}.$$

<u>Numerical examples</u>:

1) Consider

$$y' = 1 + y^2$$

$$y(0) = 1$$

with the obvious solution

$$y(x) = \tan\left(x + \frac{\pi}{4}\right) \text{ with pole at } x = \frac{\pi}{4} = 0.785\ 398\ 163.$$

We like to integrate from 0 to 0.5, where we have the exact value

$$y(0.5) = \tan(73^o,647\ 889\ 76)$$

$$= 3.408\ 223\ 442.$$

With fixed $\alpha = -1$ and the step size h we obtain

h	$u(0.5,h)$	$w(0.5,h)$	$w(0.5,h)/h^4$
1/8	3.408 278 84	$5.54.10^{-5}$	0.227
1/16	3.408 226 95	$3.50.10^{-6}$	0.229
1/32	3.408 223 66	$2.13.10^{-7}$	0.223

For cubic splines (i.e., a linear method) we would find

1/32	3.408 333 04	$1.10.10^{-4}$.

We used $w(x,h) = u(x,h) - y(x)$ to denote the error. The last column showed the error normalised by dividing with h^4. It is apparent that the error is almost proportional to h^4 - the method seems to have a *global error* of *the 4th order.*

2) Consider $y' = 1 + y^2 + y^4$

$$y(0) = 1.$$

We integrate until the spline becomes singular, indicating the solution should be singular too. The step size is kept constant, control is by testing whether $(1 + \frac{z}{b_j})$ will vanish between x_j and x_{j+1}, and $-b_j + x_j$ provides the quoted estimates for the location of the singularity.

Exact location of singularity:

$$x_s = \frac{\pi\sqrt{3}}{12} - \frac{\ln 3}{4} = 0.178\ 797.$$

For variable exponents α we find

h	$\tilde{x}_s$
2^{-5}	0.177 36
2^{-6}	0.178 61
2^{-7}	0.178 68
2^{-8}	0.178 77.

For fixed exponents α we get the following table, that shows that $\alpha = -1/3 = \frac{1}{1-4}$ is the best choice.

$h\backslash\alpha$	$-\frac{1}{10}$	$-\frac{1}{5}$	$-\frac{1}{3}$	$-\frac{1}{2}$	-1
2^{-5}	0.175 47	0.176 98	0.179 38	0.181 56	0.189 27
2^{-6}	0.177 64	0.178 15	0.178 835	0.179 70	0.182 34
2^{-7}	0.177 80	0.178 24	0.178 825	0.179 56	0.181 79.

3) An example in which the exact solution is not known:

$y' = y^4 + x^2 \cdot y^6$

$y(0) = 1$.

Again we estimate the location $\tilde{x}_s$ of the singularity of the solution. Consider first splines with fixed exponent

$h\backslash\alpha$	$-\frac{1}{10}$	$-\frac{1}{5}$	$-\frac{1}{3}$	$-\frac{1}{2}$	-1
2^{-5}	0.299 50	0.300 96	0.302 96	0.305 37	0.312 82
2^{-6}	0.300 68	0.301 04	0.301 51	0.302 09	0.303 83
2^{-7}	0.300 714	0.301 03	0.301 47	0.303 19	0.303 65

α variable:

h	$\tilde{x}_s$
2^{-4}	0.301 72
2^{-5}	0.301 45
2^{-6}	0.301 08
2^{-7}	0.301 05.

Obviously α = -1/5 is optimal.

Observe the monotonicity of the estimated $\tilde{x}_s$ in dependence upon α, if α is kept fixed, in examples 2 and 3.

§11 Convergence of Spline Method for the Solution of the Initial Value Problem

Our examples showed that it is convenient to parametrize u by means of its derivatives up to the $(k+1)$st order at the left hand endpoint of the subinterval. In many examples (as in the case of rational splines) one may use $u^{(k+1)}(x_j)$ or $u^{(k)}(x_{j+1})$ as the $(k+2)$nd parameter.

We assume that

$u(z;x,u,\ldots,u^{(k+1)})$ be defined on a domain

$$|z| \le H_0, x_0 \le x \le x_+, (x,u,\ldots,u^{(k+1)}) \in L \subset \mathbb{R}^{k+3}$$

and $k+3$ times differentiable, with

$$\text{(11.1)}\qquad \frac{d^v}{dz^v} u(z;x,u,\ldots,u^{(k+1)})\Big|_{z=0} = u^{(v)}(x) \quad \text{for } v = 0,\ldots,k+1.$$

We first investigate the solvability of equation (10.4). Since we will use the contracting mapping theorem in some special form we first formulate the following simple

Lemma 1.1:

1) Let $\phi(v,h)$ be a realvalued function defined for $|v - v_0| \le \eta$, v_0 fixed, and $0 \le h \le h_0$, continuous in h and contracting in v with coefficient $K < 1$.

2) Let $\phi(v_0,0) = v_0$.

Then there is $h^* \in (0,H_0]$ such that for every $h \in [0,h^*]$ the set $|v - v_0| \le \eta$ is mapped into itself by ϕ; hence there is a fixed point $v(h)$ such that

$$\text{(11.2)}\qquad \phi(v(h),h) = v(h) \quad \forall\, h \in [0,h^*].$$

Obviously $v(h)$ is continuous in h.

Proof: Let $\varepsilon = (1-K)\eta$ and find h^* such that

$$|\phi(v_0,h) - \phi(v_0,0)| < \varepsilon \text{ for every } h \in (0,h^*]$$

holds. For h restricted in this way we have for

$w = \phi(v,h)$ the inequalities

$$|w - v_0| \le |\phi(v,h) - \phi(v_0,h)| + |\phi(v_0,h) - \phi(v_0,0)|$$

$$\le K|v - v_0| + \varepsilon \le K\cdot\eta + (1-K)\cdot\eta = \eta.$$

This verifies that $|v - v_0| \le \eta$ is mapped into itself. The other statements about existence and continuous dependence of the fixpoint are standard in numerical analysis (compare e.g., Werner [18]). □

Now we are able to formulate conditions for the solvability of (10.4), restricting attention to the scalar case.

Theorem 11.1: Assumptions:

1) Suppose $f(x,y) \in C^3(G)$, $G \subset \mathbb{R}^2$.

2) Let the initial value problem P_3 have a solution in $[x_0,x_+]$, assume $(x,y(x),y'(x),y''(x),y'''(x)) \in L$ for $x_0 \le x \le x_+$.

3) For given $\delta^1 > 0$ let $C_0(h), C_1(h), C_2(h)$ be continuous functions s.t. $|C_0(h)| \le \delta^1, |C_1(h)| \le \delta^1, |C_2(h)| \le \delta^1$ for $0 \le h \le h_0$. Assume

$$(11.3) \quad \begin{cases} u_0(h) = y_0 + h^4\cdot C_0(h) & (y_0 = y(x_0),\ \text{etc.}) \\ u_0'(h) = y_0' + h^4\cdot C_1(h) \\ u_0''(h) = y_0'' + h^2\cdot C_2(h) \end{cases}$$

and $u_0''' \in [y_0''' - \delta, y_0''' + \delta]$ to be such that

$$(x_0, u_0(h), u_0'(h), u_0''(h), u_0''') \in L \quad \text{for } 0 \le h \le h_0$$

and every admissible u_0'''.

Then there is an $h^* \in (0,h_0)$ such that for every $h \in [0,h^*]$ there is a segment $u(x,h)$ of a spline which is defined in $[x_0,x_0 + h]$ having the initial data (11.3) and satisfying (10.4), i.e.,

$$(11.4) \quad u'(x_0+ h,h) = f(x_0 + h,u(x_0 + h,h))$$

and

(11.5) $u_0'''(h) = u'''(x_0,h) = y_0''' + h\cdot C_3(h)$,

where $C_3(h)$ is a continuous function.

Proof: We will rewrite equation (11.4) as fixed point equation for the function $C_3(h)$ and apply Lemma 11.1.

Let

(11.6) $$v_0 = \frac{1}{3}[y^{(4)}(x_0) - u^{(4)}(0;x_0,y_0,y_0',y_0'',y_0''')] - 2C_2(0).$$

Then we will choose η appropriately and use

$|v - v_0| \leq \eta$ to define

$\phi(v,h) = v_0$ for $h = 0$,

(11.7) $$\phi(v,h) = \Big[f(x_0 + h,u(h,x_0,u_0(h),u_0'(h),u_0''(h),y_0''' + hv)) - u_0'(h) - u_0''(h)h - y_0'''\frac{h^2}{2} - \frac{h^3}{3!}A(h,hv)\Big]\frac{2}{h^3} \text{ for } h \neq 0,$$

where A denotes the error term of the Taylor expansion of $u'(x_0 + h,h)$ at x_0, say by an integral whose integrand contains the factor $u^{(4)}(\tau,x_0,u_0(h),u_0'(h),u_0''(h),y_0''' + hv)$ analogously to the expansion (6.2) with $f = u'$, i.e.

$$A(h,hv) = 3\int_0^1 t^2\cdot u^{(4)}(h - ht,x_0,u_0,u_0',u_0'',y_0''' + v(h))dt.$$

We may choose η such that $\eta\cdot h_0 \leq \delta$ holds; hence by assumption 3; the arguments $(x_0,u_0(h),u_0'(h),u_0''(h),y_0''' + hv)$ for $|v| \leq \eta$, $h \in [0,h_0]$ are admissible for u to yield $u(x,h) = u(x,x_0,u_0(h), \ldots,y_0''' + hv)$. Next we have to verify *continuity with respect to h* for fixed v. For $h \neq 0$ continuity of $\phi(v,h)$ is obvious. To analyse $h \to 0$ we observe, that by Taylor expansion about x_0 for u and its first derivative,

(11.8) $$u(x_0 + h,h) - y(x_0 + h) = h^4\cdot C_0(h) + h^5\cdot C_1(h) + \frac{h^4}{2!}C_2(h) + \frac{h^4}{3!}v + \frac{h^4}{3!}\int_0^1 t^3[u^{(4)} - y^{(4)}]dt$$

and

$$(11.9) \qquad u_0'(h) + u_0''(h)h + y_0''' \frac{h^2}{2} + \frac{h^3}{3!} A(h,hv) - y'(x_0 + h)$$

$$= h^4 \cdot C_1(h) + h^3 \cdot C_2(h) + \frac{h^3}{3!}(A - y^{(4)}(x^*))$$

with an appropriate point x^*. Therefore

$$(11.10) \qquad \phi(v,h) = \frac{2}{h^3}\left[f(x_0 + h, y(x_0 + h) + O(h^4)) - y'(x_0 + h)\right]$$

$$- \left[C_2(h) + \frac{1}{3!}(A - y^{(4)}(x^*))\right] \cdot 2$$

$$= 2 \cdot f_y(\ldots) \cdot O(h^{4-3}) - 2\left[C_2(h) + \frac{1}{3!}(A - y^{(4)}(x^*))\right]$$

$$\to v_0 \text{ for } h \to 0.$$

It remains to verify that $\phi(v,h)$ is *contracting* in v and, since $\phi(v,0)$ is independent of v, it is sufficient to consider $h \neq 0$. Due to the assumptions on $u(z,x,\ldots)$ and the analysis of the remainder terms of Taylor expansions, A is differentiable with respect to u_0''' and since $u_0''' = y''' + hv$ we even gain an additional factor h. Also we obtain

$$\frac{\partial}{\partial v} f(x_0 + h, u(x_0 + h, x_0, \ldots, y_0''' + hv) \quad = f_y \cdot u_v$$

and, because of (11.8),

$$= f_y \cdot \left[\frac{h^4}{3!} + \frac{h^4}{3!} \cdot \int_0^1 t^3 \frac{\partial}{\partial v} u^{(4)}(x_0 + h - ht, x_0, \ldots, y_0''' + hv) dt\right].$$

Therefore

$$|\phi(v,h) - \phi(w,h)| \le K \cdot |v - w|,$$

where

$$(11.11) \qquad K = h\left[\|f_y\| \cdot \left(\frac{1}{3} + \frac{h}{12}\|u^{(4)}_{u'''}\|\right) + \frac{1}{3}\|u^{(4)}_{u'''}\|\right]$$

may be used. By a possible decrease of the bound h_0 we achieve

$K < 1$. Hence all assumptions of the lemma are satisfied and we have as a consequence the existence of $h^* \le h_0$ and of a fixed point $v(h)$ for $0 \le h \le h^*$ as stated; the quantity h^* depends only on δ^1, the choice of δ and K. With $C_3(h) = v(h)$ the theorem is proved. □

Remarks: 1) The domain defined by (11.3) also guarantees uniform bounds for the higher order derivatives of u due to the boundedness of $u_0, \ldots, u_0'''$. Hence we may get now a higher order expansion of $C_3(h)$ with respect to h.

2) Instead of constructing the spline on $[x_0, x_0 + h]$, we could use any $x^* \in (x_0, x_+)$ and data (11.3) given at this point to construct splines in $[x^*, x^* + h]$ for h sufficiently small.

The bound h^* for h could be chosen uniformly with respect to L apart from the fact that

$$x^* + h^* \le x_+$$

should hold.

For the time being we assume remark 1 to get

Corollary 11.1: Suppose the assumptions of Theorem 11.1 are satisfied and in addition

$$f(x,y) \in C^{(4)}(G)$$

and

$$C_2(h) = C_2 + h \cdot C_2^*(h), \quad C_2^*(h) \text{ continuous in } 0 \le h \le h_0.$$

Then the third derivative of $u(z, x, u_0(h), \ldots, u_0'''(h))$, which solves (11.3) and (11.4), admits an expansion

$$(11.12) \qquad u_0'''(h) = y_0''' + h \cdot C_3 + h^2 \cdot C_3^*(h)$$

with

$$(11.13) \qquad \begin{cases} C_3 = \frac{1}{3}\left[y^{(4)}(x_0) - u^{(4)}(0; x_0, y_0, \ldots, y_0''')\right] - 2C_2 \text{ and} \\ C_3^*(0) = \frac{1}{12}\left[y^{(5)}(x_0) - u^{(5)}(0, x_0, y_0, \ldots, y_0''') \right. \\ \qquad \left. - \frac{\partial}{\partial y_0'''} u^{(4)}(0, x_0, y_0, \ldots, y_0''') \cdot C_3\right] \end{cases}$$

$$\left| \begin{array}{l} - 2C_1 - 2C_2^*(0) + 2f_y(x_0,y_0)\cdot C^*, \\ C^* = C_0 + \frac{C_2}{2} + \frac{C_3}{6} + \frac{1}{24}(u^{(4)}(0,x_0,y_0,\ldots,y_0''') - y^{(4)}(x_0)). \end{array} \right.$$

Proof: By (11.3), (11.4), (11.5) and (11.7) we have

$$C_3(h) = C_3 + h\cdot C_3^*(h) = v(h) = \phi(v(h),h);$$

therefore it suffices to expand $\phi(v(h),h)$ as given in (11.7) with respect to h. Let $v = v(h)$. We use

$$u(h,x_0,u_0(h),\ldots,y_0''' + h\cdot v) = y(x_0 + h) + h^4\cdot C_0(h)$$

$$+ h^5\cdot C_1(h) + \frac{h^4}{2} C_2(h) + \frac{h^4}{3!} v$$

$$+ \frac{h^4}{3!}\int_0^1 t^3[u^{(4)} - y^{(4)}]dt$$

and

$$f(x_0 + h,y(x_0 + h)) = y'(x_0 + h)$$

to rewrite ϕ and find

$$\begin{aligned} (11.14)\qquad \frac{h^3}{2}\phi(v,h) &= y'(x_0 + h) - u_0' - u_0''h - y_0'''\frac{h^2}{2} \\ &- \frac{h^3}{2}\int_0^1 t^2 u^{(4)}(\ldots)dt + f_y(x_0 + h,\tilde{u})\cdot h^4 \\ &\cdot\left[C_0(h) + \frac{C_2(h)}{2} + \frac{1}{3!}\int_0^1 t^3[u^{(4)} - y^{(4)}]dt\right. \\ &\left. + \frac{v}{3!} + hC_1(h)\right] = - h^4\cdot C_1(h) - h^3\cdot(C_3 + h\cdot C_3^*(h)) \\ &- \frac{h^3}{2}\int_0^1 t^2\left[u^{(4)}\Big|_{h=0} + h\frac{d}{dh}u^{(4)}\Big|_{h=0} - y^{(4)}\Big|_{h=0}\right. \\ &\left. - y^{(5)}\Big|_{h=0}(1 - t)\right]dt + h^4\cdot f_y\cdot[\ldots] \end{aligned}$$

$$= -h^3 \cdot C_3 - \frac{h^3}{2} \cdot \int_0^1 t^2 [u^{(4)} - y^{(4)}]\Big|_{h=0} dt$$

$$- h^4 \cdot C_1 - h^4 \cdot C_3^* - \frac{h^4}{2} \cdot \int_0^1 t^2 (1-t) [u^{(5)}$$

$$- y^{(5)}]\Big|_{h=0} dt - \frac{h^4}{2} \cdot \int_0^1 t^2 \cdot \frac{\partial}{\partial u'''} u^{(4)} (0, x_0, y_0,$$

$$\ldots, y_0''') C_3 dt + h^4 \cdot f_y [\ldots] + 0(h^5).$$

Observing $\int_0^1 t^2 dt = \frac{1}{3}$, $\int_0^1 t^2 (1-t) dt = \frac{1}{12}$ the stated result is obtained. □

We point out that C_3^*, just as C_3, is determined by the higher order derivatives of u, y and $f(x,y)$.

As the next step we have to estimate the error incurred in $[x_0, x_1]$, i.e., the magnitude of

(11.15) $\quad w(x,h) = u(z, x_0, u_0(h), \ldots, u_0'''(h)) - y(x), \quad z = x - x_0$

and its derivatives, in order to obtain the initial data needed at x_1 for the determination of u on $[x_1, x_2]$.

We assume $C_i(h) = C_i(0) + h \cdot C_i^*(h)$ to hold.

We will obtain formal expansions and observe that the initial data found to hold at x_1 will stay in the domain of the parameters of u, if h is appropriately restricted.

We use the Taylor expansion of w and its derivatives, observe the initial conditions given by (11.3) and the result (11.12) and (11.13) to obtain the following equations.

(11.16) $\quad w_1 = w(x_1, h) = w_0 + w_0' \cdot h + w_0'' \cdot \frac{h^2}{2} + w_0''' \cdot \frac{h^3}{3!}$

$$+ w_0^{(4)} \frac{h^4}{4!} + \frac{h^5}{4!} \cdot \int_0^1 t^4 \cdot w^{(5)} dt$$

$$= h^4\cdot\{C_0 + C_1\cdot h + C_2\cdot\frac{1}{2} + C_3\cdot\frac{1}{3!} + C_4\cdot\frac{1}{4!} + 0(h)\}$$

$$= h^4\cdot\{C_0 + C_2\cdot\frac{1}{2} + [-2C_2 - \frac{1}{3}\cdot C_4]\cdot\frac{1}{3!} + C_4\cdot\frac{1}{4!}\} + 0(h^5)$$

$$= h^4\cdot\{C_0 + \frac{1}{6}C_2 - \frac{1}{72}C_4\} + 0(h^5)$$

where $C_4 = u^{(4)}(x_0,\ldots) - y^{(4)}(x_0) = w^{(4)}(x_0 + 0,h)$ is used. Hence

$$(11.17)\qquad w_1' = w'(x_1,h) = w_0' + w_0''h + w_0'''\frac{h^2}{2} + w_0^{(4)}\cdot\frac{h^3}{3!} + \frac{h^4}{3!}\cdot\int_0^1 t^3 w^{(5)}\,dt$$

$$= C_1h^4 + \underbrace{h^3(C_2 + C_3\cdot\frac{1}{2} + C_4\cdot\frac{1}{6})}_{=\,0(h).} + 0(h^4) = 0(h^4),$$

Analogously

$$(11.18)\qquad w_1'' = h^2\cdot(C_2 + [-2C_2 - \frac{1}{3}C_4] + \frac{1}{2}C_4) + 0(h^3)$$

$$= h^2\cdot(-C_2 + \frac{1}{6}C_4) + 0(h^3)$$

$$(11.19)\qquad w'''(x_1 - 0,h) = h(-2C_2 + \frac{2}{3}C_4) + 0(h^2)$$

$$(11.20)\qquad w^{(4)}(x_1 - 0,h) = C_4 + 0(h).$$

It is long but not difficult to find the explicit expressions of the $0(h^p)$ - terms. It is clear, however, that the factors of h^p are controlled (bounded) by a barrier depending upon $u_0,\ldots,u_0'''$, hence uniformly bounded as long as

$$(x_0,u_0,\ldots,u_0''') \in K$$

where we define

$$(11.21)\qquad K = \{(x,u,\ldots,u''')\,\Big|\, x_- \le x \le x_+, |u - y(x)| \le \delta,$$

$$|u' - y'(x)| \le \delta, |u'' - y''(x)| \le \delta,$$

$$|u''' - y'''(x)| \le \delta\}$$

and δ is chosen such that the pointset

$$\{(z,x,u,\ldots,u''') \,\Big|\, |z| \le H_0,\ (x,\ldots,u''') \in K\} \subset L.$$

L is the domain in which u is defined.

In $[x_1,x_2]$ the spline may now be defined. In particular

(11.22) $\quad u_1^{(i)} = y_1^{(i)} + h^{4-i} C_{1i}(h) \quad$ for $i = 0,\ldots,4,$

where in case $i = 3,4$ the limits from the right are described. In fitting the spline in $[x_1,x_2]$ as before the relation

(11.23) $\quad C_{13} = -2C_{12} - \frac{1}{3} C_{14}$ is obtained from (11.13).

Since w, w', w'' are continuous, u, u', u'' vary only slightly from x_0 to x_1 and by previous continuity consideration C_{13} will differ only slightly from C_3 if h is small; hence

(11.24) $\quad u^{(4)}(0;x_1,u_1(h),u_1'(h),u_1''(h),y_1''' + h\cdot C_{13}(h))$

$$= u^{(4)}(0,x_0,u_0(h),u_0'(h),u_0''(h),u_0'''(h)) + 0(h).$$

The fourth order derivative is not (necessarily) continuous at x_1 but by (11.24) and (11.20) the *jump is of order h*.

It is also interesting to calculate the jump of the third order derivatives. We express C_{12} and C_{14} by (11.24), (11.18) and use (11.19) to find

(11.25) $\quad w'''(x_1 + 0,h) - w'''(x_1 - 0,h) = h\cdot C_{13} - h(-2C_2 + \frac{2}{3} C_4)$

$$+ 0(h^2) = h\cdot(2C_2 - \frac{2}{3} C_4)\cdot 2 + 0(h^2).$$

(11.26) $\quad w'''(x_1 + 0,h) + w'''(x_1 - 0,h) = 0(h^2),$

i.e., the jump of u''' at x_1 is of the order h, the average $\frac{1}{2}(u'''(x_1 + 0,h) + u'''(x_1 - 0,h))$ differs from $y'''(x_1)$ by $0(h^2)$.

We have thus found a good picture of the local behaviour of the error: $w^{(i)}(x,h) = 0(h^{4-i})$ for $i = 0,\ldots,4$ in $[x_0,x_1]$.

To investigate the error propagation and the global order of convergence denote

$$(11.27)\qquad \overline{c}_j = \begin{pmatrix} c_{j0} \\ c_{j2} \\ c_{j4} \end{pmatrix}, \quad \text{for } j > 0,\ \overline{c}_0 = \begin{pmatrix} c_0 \\ c_2 \\ c_4 \end{pmatrix}.$$

Then we may summarize equations (11.16) - (11.20) by

$$(11.28)\qquad \overline{c}_1 = A\cdot\overline{c}_0, \quad A = \begin{pmatrix} 1 & \frac{1}{6} & -\frac{1}{72} \\ 0 & -1 & \frac{1}{6} \\ 0 & 0 & 1 \end{pmatrix}$$

as far as the leading coefficients are concerned.

With the same arguments as before we may investigate the leading coefficients of the error at x_2, to find

$$\overline{c}_2 = A\cdot\overline{c}_1 = A^2\cdot\overline{c}_0 = \overline{c}_0, \quad \text{since } A^2 = E = \begin{pmatrix} 1 & 0 & 0 \\ 0 & 1 & 0 \\ 0 & 0 & 1 \end{pmatrix}.$$

The previous analysis will hold at any point x, if the parameters $u,\ldots,u'''$ are such that they lead to a point of K.

Let us assume all parameters which we use in the following satisfy this condition - it will be guaranteed by choosing the h-interval sufficiently small.

Then we may denote

$$(11.29)\qquad \overline{v}_j = \begin{pmatrix} w_j/h^4 \\ w_j''/h^2 \\ w_j^{(4)} \end{pmatrix}$$

such that

$$(11.30)\qquad \overline{v}_{j+1} = (A + h\cdot C_j^*)\cdot\overline{v}_j + 0(h),$$

where bounds for C_j^* could be found that hold uniformly in K.

Since

$$(A + h\cdot C_j^*)(A + h\cdot C_{j-1}^*) = E + h\cdot K_j$$

with $\|K_j\| \le C$, we can use the well known considerations of single step methods to conclude that

(11.31) $\quad |\overline{v}_j| \le \text{const} \cdot |\overline{v}_0| + \text{const},$

(actually one could (and should) also keep track of $w'(x,h)$) hence the errors $w^{(i)}(x,h)$ uniformly in h satisfy

(11.32) $\quad w^{(i)}(x,h) = 0(h^{4-i})$

provided $y^{(i)}(x) + w^{(i)}(x,h)$ make up admissible parameters, lying in K. But this is just achieved, if $|\overline{v}|_j \cdot h_0 \le \delta$, which is just a matter of limiting h, say, by $[0,h^*]$. This sketch shows how fourth order convergence is obtained.

§12. Numerical Stability of Spline Method

The investigations of the last paragraph showed that the algorithm for the solution of initial value problems (see §10) yields fourth order convergence in case $k = 2$. One could hope to get even better convergence if $k > 2$.

As observed in numerical experience, there is a natural obstacle, the numerical stability of the method. We will indicate where the difficulties arise, while we will not bother with setting up rigorous domains etc.

The matrix A defined in (11.28) has eigenvalues of absolute value equal to 1, therefore the method does not intrinsically enlarge the propagated round-off errors. Indeed it could be verified to be a non-linear weakly stable two step method.

Assuming $k > 2$, we will derive the corresponding matrix A_k. For this method the values $u(x_0),\ldots,u^{(k)}(x_0)$ shall be specified as initial data while $u^{(k+1)}(x_0)$ has been chosen to satisfy equation (10.4). Then we could expand

$$w(x,h) = u(x,h) - y(x)$$

and its derivatives as before to find the local behaviour of the error.

Let $w^{(i)}(x,h) = h^{k-i+2}\cdot C_i(h)$ analogous to (11.3) and $C_i = C_i(0)$. Then (10.4) gives the equation $C_1(0) = 0$ and

$$(12.1) \qquad \frac{1}{1!}C_2 + \frac{1}{2!}C_3 + \dots + \frac{1}{k!}C_{k+1} + \frac{1}{(k+1)!}C_{k+2} = 0,$$

which determines

$$(12.2) \qquad C_{k+1} = -k!\left[\frac{1}{1!}C_2 + \dots + \frac{1}{(k-1)!}C_k + \frac{1}{(k+1)!}C_{k+2}\right].$$

The Taylor expansions are

$$(12.3) \qquad w^{(i)}(x_1,h) = \sum_{j=0}^{k+2-i} w_0^{(i+j)}\cdot\frac{h^j}{j!} + 0(h^{k+3-i}), \quad 0 \le i \le k+2,$$

which we had used already to find (12.1) for $i = 1$. Now we may employ it for the other values of i to get the transition for the leading coefficients of the error.

$$(12.4) \qquad C_k = \sum_{0}^{k+2-i} C_{j+i}\cdot\frac{1}{j!} \quad \text{(singling out } C_{k+1}\text{)}$$

$$= \sum_{0}^{k-i} C_{i+j}\cdot\frac{1}{j!} + C_{k+2}\cdot\frac{1}{(k+2-i)!}$$

$$- \frac{k!}{(k+1-i)!}\left[\sum_{j=1}^{k-1} C_{1+j}\,\frac{1}{j!} + C_{k+2}\cdot\frac{1}{(k+1)!}\right]$$

$$= \sum_{j=1}^{k-1} C_{1+j}\left[-\frac{k!}{(j!)(k+1-i)!} + \frac{1}{(j-i+1)!}\right]$$

$$+ C_{k+2}\,\frac{1}{(k+2-i)!}\cdot\frac{i-1}{k+1}$$

for $i = 2,\dots,k$ and with $\frac{1}{v!} = 0$ if $v < 0$.

Again we could combine the coefficients C_j to form

$$\overline{C}_j = \begin{pmatrix} C_{j0} \\ C_{j2} \\ \vdots \\ C_{jk} \\ C_{jk+2} \end{pmatrix}, \quad \overline{C}_0 = \begin{pmatrix} C_0 \\ C_2 \\ \vdots \\ C_k \\ C_{k+2} \end{pmatrix}.$$

The coefficients of (12.4) then describe the elements a_{ij} of A_k of the second through kth row.

$$\overline{C}_{j+1} = A_k \cdot \overline{C}_j .$$

Obviously the first and last of the diagonal elements are equal to 1 as before. The other $k - 1$ diagonal elements are of the form ($i = 2 + j$ in (12.4))

(12.5) $$d_j = \frac{1}{0!} - \frac{k!}{j!(k-j)!} \quad \text{for } j = 1,\dots,k-1.$$

Since

$$\binom{k}{j} = \frac{k!}{j!(k-j)!} > 2 \quad \text{if } k > 2,\ 1 \le j \le k-1,$$

we see that the sum of these $k - 1$ diagonal elements is

$$\sum_{j=0}^{k-2} d_j = k - 1 - \sum_{j=0}^{k-2} \binom{k}{1+j} < 1 - k.$$

Therefore the $(k - 1) \times (k - 1)$ matrix obtained from A_k by deleting the first row and column (which is equal $\begin{pmatrix} 1 \\ 0 \\ \vdots \\ 0 \end{pmatrix}$) and last row (equal $(0,\dots,0,1)$) and column has eigenvalues with sum less than $(1 - k)$, i.e., there is at least one eigenvalue of absolute value larger than 1. By the indicated special form of A_k this implies that A_k has an eigenvalue λ with $|\lambda| > 1$. This explains the instability in propagating round off errors.

For the convenience of the reader we remark that

$$A_3 = \begin{pmatrix} 1 & \frac{1}{4} & \frac{1}{24} & -\frac{1}{480} \\ 0 & -2 & -\frac{1}{2} & \frac{1}{24} \\ 0 & -6 & -2 & \frac{1}{4} \\ 0 & 0 & 0 & 1 \end{pmatrix}$$

with the eigenvalues $1,1,-2+\sqrt{3},-2-\sqrt{3}$.

To get stable algorithms one could look for methods that would not only match u' by (10.4) but also u'',... at the right hand side of the interval $[x_j, x_{j+1}]$, reducing the number of initial values (10.3) correspondingly, but we cannot pursue these considerations in this paper.

REFERENCES

[1] Arndt, H., Interpolation mit regulären Spline-Funktionen, Dissertation, Münster, 1974.

[2] Baumeister, J., Extremaleigenschaften nichtlinearer Splines, Dissertation, München, 1974.

[3] Braess, D. and Werner, H., Tschebyscheff-Approximation mit einer Klasse rationaler Splinefunktionen II, J. Approximation Theory 10, 379-399, (1974).

[4] Lambert, J.D. and Shaw, B., On the numerical solution of $y' = f(x,y)$ by a class of formulae based on rational approximation, Math. Comp. 19, 456-462, (1965).

[5] Lambert, J.D. and Shaw, B., A method for the numerical solution of $y' = f(x,y)$ based on a self-adjusting non-polynomial interpolant, Math. Comp. 20, 11-20, (1966).

[6] Loscalzo, F.R. and Talbot, T.D., Spline function approximation for solutions of ordinary differential equations, SIAM J. Num. Anal. 4, 433-445, (1967).

[7] Micula, G., Bemerkungen zur numerischen Lösung von Anfangswert-problemen mit Hilfe nichtlinearer Splinefunktionen, In Lecture Notes in Mathematics 501, 200-209, Spline Functions, Karlsruhe 1975, Springer Verlag, Berlin-Heidelberg-New York, (1976).

[8] Runge, R., Lösung von Anfangswertproblemen mit Hilfe nichtlinearer Klassen von Spline-Funktionen, (Dissertation Münster, 1972).

[9] Schaback, R., Spezielle rationale Splinefunktionen, J. Approximation Theory 7, 281-292, (1973). (Dissertation Münster, 1969).

[10] Schaback, R., Interpolation mit nichtlinearen Klassen von Splinefunktionen, J. Approximation Theory 8, 173-188, (1973).

[11] Schomberg, H., Tschebyscheff-Approximation durch rationale Splinefunktionen mit freien Knoten, Dissertation, Münster, 1973.

[12] Werner, H., Tschebyscheff-Approximation mit einer Klasse rationaler Splinefunktionen, J. Approximation Theory 10, 74-92, (1974).

[13] Werner, H., Tschebyscheff-Approximation nichtlinearer Splinefunktionen, in K. Böhmer-G. Meinardus-W. Schempp Spline-Funktionen, BI-Verlag Mannheim-Wien-Zürich, 303-313, (1974).

[14] Werner, H., Interpolation and Integration of Initial Value Problems of Ordinary Differential Equations by Regular Splines, SIAM J. Num. Anal. 12, 255-271, (1975).

[15] Werner, H., Numerische Behandlung gewöhnlicher Differentialgleichungen mit Hilfe von Splinefunktionen, ISNM 32, 167-175, (1976).

[16] Werner, H., Approximation by Regular Splines with Free Knots, Austin, Symposium on Approximation Theory 1976, S. 567-573.

[17] Werner, H. and Loeb, H., Tschebyscheff-Approximation by Regular Splines with Free Knots, in Approximation Theory, Bonn 1976, Lecture Notes in Math. 556, 439-452, Springer Verlag, Berlin-Heidelberg-New York (1976).

[18] Werner, H., Praktische Mathematik I, 2. Auflage, Springer-Verlag, Berlin-Heidelberg-New York 1976.

[19] Werner, H. and Schaback, R., Praktische Mathematik II, 2. Auflage, Springer-Verlag, Berlin-Heidelberg-New York 1978.

[20] Werner, H. and Wuytack, L., Nonlinear Quadrature Rules in the Presence of a Singularity, Universiteit Antwerpen, to appear in CAMWA 1978.

[21] Werner, H. and Zwick, D., Algorithms for Numerical Integration with Regular Splines, Rechenzentrum der Universität Münster Schriftenreihe Nr. 27, (1977).

[22] Wuytack, L., Numerical Integration by Using Nonlinear Techniques, J. of Comp. and Appl. Math. 1, 267-272, (1975).

LIST OF PARTICIPANTS

Barnes, D.C.	Mathematics Dept., Washington State University, Pullman, Washington 99163, USA.
Bogar, G.A.	Dept. of Mathematics, Montana State University, Bozeman, Montana 59717, USA.
Boivin, A.	4800 Beaugrand, Montreal, Quebec, H1K 3Y5, Canada.
Chui, C.K.	Dept. of Mathematics, Texas A & M University, College Station, Texas 77843, USA.
Collatz, L.	Inst. für Angewandte Mathematik, Universität Hamburg, Bundesstr. 55, 2 Hamburg 13, Germany.
Dixon, M.J.	Dept. of Mathematics, California State University, Chico, California 95929, USA.
Dougalis, V.A.	Dept. of Mathematics, University of Tennessee, Knoxville, Tennessee 37916, USA.
Drew, D.M.	Dept. of Mathematics, Brunel University, Uxbridge, Middlesex, UB8 3PH, UK.
Eggert, N.H.	Dept. of Mathematics, Montana State University, Bozeman, Montana 59717, USA.
Fournier, R.	Déptt. de Mathématiques et de statistiques, Université de Montréal, C.P.6128, Montréal, Québec H3C 3J7, Canada.

Garder, A.	Dept. of Mathematics, Southern Illinois University, Edwardsville, Illinois 62026, USA.
Gribble, J.D.	Dept. of Applied Mathematics, University of St. Andrews, Fife, Scotland KY16 9SS, UK.
Hall, C.A.	Inst. for Computational Mathematics & Applications, Dept.of Mathematics and Statistics, University of Pittsburgh, Pittsburgh, Pennsylvania 15260, USA.
Henry, M.S.	Dept. of Mathematics, Montana State University, Bozeman, Montana 59717, USA.
Holland, A.S.B.	Dept. of Mathematics & Statistics, University of Calgary, 2920-24th Avenue, N.W., Calgary, Alberta T2N 1N4, Canada.
King, J.P.	Dept. of Mathematics, Lehigh University, Bethlehem, Pennsylvania 18015, USA.
Kramarz, L.	Dept. of Mathematics, Emory University, Atlanta, Georgia 30322, USA.
Kratz, L.J.	Dept. of Mathematics, Idaho State University, Pocatello, Idaho 83209,USA.
Lancaster, P.	Dept. of Mathematics & Statistics, University of Calgary, 2920-24th Avenue, N.W., Calgary, Alberta T2N 1N4, Canada.
Lavoie, M.	Université du Québec-Rimouski, 300 ave. des Ursulines, Rimouski, Québec G5L 3A1, Canada.
Leeming, D.J.	Dept. of Mathematics, University of Victoria, P.O.Box 1700, Victoria, British Columbia V8W 2Y2, Canada.
McCabe, J.H.	Mathematical Institute, University of St. Andrews, St. Andrews, Fife, Scotland KY16 9SS, UK.
McNab, G.S.	Imperial Oil Ltd (Production Dept.), 339-50 Avenue S.E., Calgary, Alberta T2G 2B3.

Meinardus, G.	Lehrstuhl IV für Mathematik, Gesamthochschule Siegen, 5900 Siegen 21, Holderlinstraße 3, West Germany.
Meinguet, J.	Université Catholique de Louvain, Inst. de Mathématique Pure et Appliquée, Chemin du Cyclotron 2, B-1348 Louvain-la-Neuve, Belgique.
Micchelli, C.A.	Mathematics Research Center, University of Wisconsin, 610 Walnut Street, Madison, Wisconsin 53706, USA.
Mohapatra, R.N.	Dept. of Mathematics, American University of Beirut, Beirut, Lebanon.
Nasim, C.	Dept. of Mathematics & Statistics, University of Calgary, 2920-24th Avenue N.W., Calgary, Alberta T2N 1N4, Canada.
Phillips, G.M.	Mathematical Institute, University of St. Andrews, St. Andrews, Fife, Scotland KY16 9SS, UK.
Puri, K.K.	Dept. of Mathematics, University of Maine, Orono, Maine 04473, USA.
Rahman, Q.I.	Dept. of Mathematics, University of Montreal, C.P.6128, Montreal, P.Q. H3C 3J7.
Rao, V.V.G.	Dept. of Mathematics, University of Cape Coast, Cape Coast, Ghana.
Rivlin, T.J.	IBM Corporation, Mathematical Sciences Dept., P.O. Box 218, Yorktown Heights, New York 10598, USA.
Sahney, B.N.	Dept. of Mathematics & Statistics, University of Calgary, 2920-24th Avenue, N.W., Calgary, Alberta T2N 1N4, Canada.
Salkauskas, K.	Dept. of Mathematics & Statistics, University of Calgary, 2920-24th Avenue N.W., Calgary, Alberta T2N 1N4, Canada.
Schmeisser, G.	Mathematisches Inst. der Universität Erlangen-Nürnberg, Bismarckstrasse 1½, D-8520 Erlangen, West Germany.

Sharma, A.	Dept. of Mathematics, University of Alberta, Edmonton, Alberta T6G 2G1, Canada.
Smith, P.W.	Dept. of Mathematics, Texas A & M University, College Station, Texas 77843, USA.
Sommer, M.	Inst. für Angewandte Mathematik der Universität Erlangen-Nürnberg, Martensstr. 3, 8520 Erlangen, West Germany.
Strauß, H.	Inst. für Angewandte Mathematik der Universität Erlangen-Nürnberg, Martensstr. 3, 8520 Erlangen,West Germany.
Taylor, P.J.	Dept. of Computing Science, University of Stirling, Stirling, Scotland FK9 4LA.
Tipper, J.C.	Kansas Geological Survey, 1930 Avenue "A", Campus West, Lawrence, Kansas 66044, USA.
Werner, H.	Westfälische Wilhelms-Universität, Inst. für Numerische und Instrumentelle Mathematik, 4400 Münster, Roxeler Straße 64, West Germany.

SUBJECT INDEX

A

B

D

E

F

G

H

I

J

K

L

M

Q

R

S